The Practical Intellect:
Computers and Skills

ARTIFICIAL INTELLIGENCE AND SOCIETY

Series Editor: KARAMJIT S. GILL

Bo Göranzon

The Practical Intellect:

Computers and Skills

With 17 Figures

Springer-Verlag
Berlin Heidelberg GmbH

Bo Göranzon, DTech
Professor
Royal Institute of Technology
S-10044 Stockholm, Sweden

Cover illustration: Lennart Mörk, *Ramanujan*

ISBN 978-3-540-19759-1

British Library Cataloguing in Publication Data
Göranzon, Bo. 1941–
    Practical intellect: Computers and Skills
    – (Artificial Intelligence and Society Series)
    I. Title. II. Series.
    006.3
    ISBN 978-3-540-19759-1   ISBN 978-1-4471-3868-6 (eBook)
    DOI 10.1007/978-1-4471-3868-6
Library of Congress Cataloging-in-Publication Data
Göranzon, Bo. 1941–
    The practical intellect: computers and skills/Bo Göranzon,
        p. cm. – (Artificial intelligence and society)
    Includes index.
    ISBN 978-3-540-19759-1
    1. Computers. 2. Electronic data processing. I. Title.
II. Series.
QA76. G5975   1992   92-19987
006.3–dc20   CIP

Composition by Genesis Typesetting, Rochester, Kent
34/3830-543210 Printed on acid-free paper

Foreword

The intention of this book is not to add another technical work to the series of publications already available on matters connected with the relations between natural and artificial intelligence, nor to repeat the positions already well expressed in, for example, the debate between John Searle, Daniel Dennet and Hubert Dreyfus. It is an attempt to encourage philosophical reflection on dimensions of the subject that have hitherto been somewhat neglected.

This book, which explores a number of case studies, is the fifth in the series, the previous four books being:

(i) *Knowledge, Skill and Artificial Intelligence* (Bo Göranzon and Ingela Josefson (Eds.), Springer-Verlag, London, 1988)

(ii) *Artificial Intelligence, Culture and Language: On Education and Work* (Bo Göranzon and Magnus Florin (Eds.), Springer-Verlag, London, 1990)

(iii) *Dialogue and Technology: Art and Knowledge* (Bo Göranzon and Magnus Florin (Eds.), Springer-Verlag, London, 1991)

(iv) *Skill and Education: Reflection and Experience* (Bo Göranzon and Magnus Florin (Eds.), Springer-Verlag, London, 1992)

An important connection between these four books is the conference on Culture, Language and Artificial Intelligence held in Stockholm in May–June, 1988. The conference was attended by more than 300 researchers and practitioners, from over 15 countries, in the fields of technology, philosophy, the history of ideas, literature and linguistics. Contributions to the books were solicited from among those who attended the conference and from researchers involved in work related to its aim.

This conference was organized by the Swedish Center for Working Life, and was related to the Center's research project based on the Education–Work–Technology triangle of concepts. This project ran from 1977 to 1991. An overview of this research is published in the report *Skill and Technology* (Magnus Florin (Ed.), Swedish Center for Working Life, Royal Institute of Technology, 1991).

The Diderot Project also started after the Stockholm conference. This project aims to provide an international forum for the discussion of the themes of enlightenment, skill and education. The epistemology of work lies at the centre of this project. Since its aim is to raise the highly unorthodox question, 'what is it that workers with skills in fact know?' from which to create the basis for a sound, genuinely social and productive implementation of new technology in the workplace, it involves people from backgrounds as widely varied as mathematics, the history of ideas, economic history, theatre, philosophy, the crafts, literary criticism, linguistics, etc. The Diderot Project sponsors meetings and discussions on these themes in Sweden and throughout Europe with a view to widening and deepening the existing networks for the social implications of technology, and this work will lead to another international symposium in Stockholm in the autumn of 1993 and the start of a new PhD programme called *Technology and Culture*, run jointly by the University of East Anglia, Norwich and the Royal Institute of Technology, Stockholm.

As is true of our own time, the eighteenth century was a difficult and complex period. There was a reaction against conferring upon science – most of all on mathematics – an almost godly ability to explain the world. There is a document from that time, Denis Diderot's *Rameau's Nephew*, which, more than any other, problematizes the way of perceiving knowledge. In this document, Diderot describes not only how difficult it is to live by a principle, but also the confrontation between different kinds of knowledge. He sets 'knowing that' against 'knowing how' and in doing so challenges a central tenet in modern philosophy, namely that practical knowledge ultimately has a theoretical basis.

In Diderot's view, the practical and the theoretical, and the experimental and the propositional, sorts of knowledge are not only radically incompatible with one another, but are also actually in competition with one another in the social world. However, the significance of the dialogue does not end there. Diderot brilliantly juxtaposes two ways of perceiving knowledge and life itself, not simply in such a way that they are questioned, ridiculed or rejected, but in a way which shows how they problematize one another. The contradictions are not resolved. It is agreed that there is a discord that cannot be bridged. The mastery in *Rameau's Nephew* lies in the fact that Diderot does not take sides in the struggle between senses and intellect, but respects the complex and contradictory interplay between the different layers of one's own person. The interaction may be seen as an example of the paradoxical view of knowledge in the *Encyclopedia* project.

We could say that it is the epistemology of professional knowledge to which Diderot calls attention in his famous dialogue and in all his works. The epistemology of professional knowledge is – notwithstanding its predecessor – a young research area whose core may be said to be the study of the development and maintenance of professional knowledge at the level of the individual, the work group and the community. An important theme in this area of research is to illustrate the effects of information technology – the use of computers – on professional knowledge and, as a secondary theme, the conditions for the development of computers and the design of computer systems.

The design of computer systems and the study of computers and skills, in particular in terms of the effects on professional knowledge, are two separate areas of research. On the basis of the set of problems involved, the research area of epistemology of professional knowledge is by its nature interdisciplinary (i.e. technology and the humanities), and requires close cooperation between academic disciplines and between authors, artists and people actively pursuing professions in the working world.

The aim of the book, as well as of the other four books in the series, is to present the contours of this new research field of skill and technology, together with a multitude of issues that demand thorough explanation. A fruitful distinction in the research process is that between exploration and surveying. This book is written more in the spirit of exploration than of surveying. It offers the reader, in the spirit of dialogue, more questions and reflections than answers. If what it indicates is true, then much more exploration will need to be done. The new discipline it implies will take some time to emerge.

Advanced professional knowledge develops through learning to see the differences in the many facets of reality. This cannot be reduced to the level of information processing by the brain. A fundamental theme in this book is to give content to the metaphors of the 'practical intellect' and the 'inner picture'. These metaphors denote a shift in perspective: there is a departure from the traditional approach to questions related to information and technical aids. As a result, the most important aspect of information, for example, becomes the extent to which it can be accommodated in the inner picture. When making their judgements, people with professional competence apply a well-founded inner picture which they have been able to build up over time. And it is the inner picture, and not the technical aids, no matter how sophisticated they may be, that gives them certainty. In the main case study in this book (the computerization of the work of forest rangers), it is noted that the ability to calculate and the ability to make judgements are two sides of the same coin. Calculation and judgement made a single whole. Computerization severed the link between calculation and judgement. No clear line can be drawn between purely routine and more advanced operations. The case study provided insight into the relationship between well-founded experience and calculation – the mathematical model – an insight which Diderot expressed as follows: 'It is a question of calculation on the one hand, and of experience on the other. If the one is well-founded, then it must agree with the other.'

It is an unavoidable fact that far more knowledge of the practical intellect is needed and, what is more, respect is needed too, because there is more to professional knowledge than one can realize as an outsider.

Many people have made important contributions to this project. They include Professor Albert Danielsson, Industrial Economics and Organization, Royal Institute of Technology, Stockholm; Professor Allan Janik, Brenner Archive, Innsbruck; Professor Julian

Hilton, Audio-Visual Multimedia, Norwich; Mr. Magnus Florin, Dramaturg, Royal Dramatic Theatre, Stockholm; Ms. Maj-Lis Perby MEng, Work Environment Fund, Stockholm; Mr. Jan-Eric Degerblad, Doctor of Technology, Work Environment Fund, Stockholm; Mr. Peter Gullers, photographer, Stockholm; Ingela Josefson, PhD, Swedish Center for Working Life, Stockholm; Professor Tore Nordenstam, Institute of Philosophy, University of Bergen; Professor Kjell S. Johannessen, Institute of Philosophy, University of Bergen; Director Jon Cook, University of East Anglia; Professor Stephen Toulmon, Northwestern University, Chicago; and Pehr Sällström, Institute of Physics, University of Stockholm. The English translation was done by Mr. Struan Robertson, who has shown a notable capability for stimulating the progress of this work. The contributions of Peter and K. W. Gullers, the photographers, and Lennart Mork, the artist, illustrate this book.

It is a privilege to work with such professional people.

Stockholm, 1992   *Bo Göranzon*

Note

All illustrations in this book are the property of Lennart Börk, Bo Göranzon and Peter Gullers, except where otherwise stated.

Contents

The adding machine invented by Blaise Pascal (1632–1662)

Chapter 1

Computers and Knowledge: An Introduction

Tabulation – A Classic Use of Computer Technology

I present to the general public a little machine which I have invented myself and which you may use as a tool to effortlessly carry out all arithmetical operations, freeing yourself from soul-destroying work when using your calculator and your pen.[1]

With these words, Blaise Pascal unveiled his adding machine, the mechanical calculator. This machine, which he designed as early as 1642 at the time of the introduction of modern natural sciences could only perform addition and subtraction. Rather less than 200 years later Charles Babbage, the English mathematician, made an important advance in the development of the mathematical calculator. He developed the principle for the so-called difference engine, which was based on Newton's method for mathematical integration. It was left to Georg Scheutz, a Swedish lawyer and publisher, and his son Edvard, to build a working model of this machine.[2]

Babbage's invention had an important impact on the production of mathematical tables. At that time they were produced by large teams of clerks who did the calculations with pen and paper. The results were compared, computed once again and then checked numerous times before anyone dared to send a copy to the printers. Babbage realized that only a machine could eliminate the human factor and produce error-free tables.[3] No one before Babbage had managed to eliminate errors in tables. He realized how important the products of such a machine would be in the daily work of mariners, astronomers, land surveyors, insurance agents, mathematicians and many others. But Babbage's ambitions went a good deal further than the difference engine. He dreamed of inventing a mechanical analytical calculator which could be programmed to solve all mathematical problems.[4]

The first computer to be designed using modern electronics was ENIAC, developed at the University of Pennsylvania and completed in 1946. It was the mathematician John von Neumann who managed to 'use electronic components to create what Babbage had failed to create with cogwheels.[5] The need to solve military problems was the driving force behind the invention of the first computers. During World War II mathematicians, biologists and professionals from the other natural sciences were employed to develop aids for use in military decision-making. This work came to be called operations analysis, and computers played a decisive part in the development of methods in this field.[6]

Operations Analysis and Computerization

At the end of the 1960s, I was an operations analyst at the Swedish National Defense Research Institute, working on the development of computerized mathematical models. My first assignment was related to something of a classic in the field of operations analysis, what is called *Lanchester's differential equations*.[7] These equations, published in 1916, were used in military applications in World War I, and similar models were later used in a wide range of studies such as charting the effects of advertising on business competition, describing the influence of propaganda in general elections and studying biological systems under stress. Common to all these applications is that there exists an adversarial relationship between the subjects of the study and that Lanchester's differential equations are used to produce tables which describe a sequence of events over time.[8] In practice, the mathematical complexity of Lanchester's differential equations makes it impossible to produce these tables by manual calculations. A set of tables could only be produced with the help of a computer.[9] Here we see a link with Babbage's ambition, although Babbage was intent upon eliminating the errors which occurred in producing tables.

My next assignment was on a completely different kind of application, comparing different technical alternatives.[10] The important difference between this work and the first study was that there was no ready-made mathematical model available, and the systems design work involved developing a mathematical model which could be modified for use on a computer. The various technical systems to be compared had different levels of support effectiveness in a defense system. We chose to make the comparison by using so-called manual games, a method in which various scenarios of possible sequences of events are simulated.[11] The point of manual games – giving the method its name – is that an aspect of human judgment is introduced into the simulations. In this case, people with specialist knowledge made assessments of various situations which occurred in the simulations.

The manual game served to gather data for the study. The technical systems, with particular characteristics separating them from one another were represented in mathematical formulae, and quantitative measurements were used to compare them.[12] The mathematical model was developed in close co-operation with a group of experts representing different areas of specialist competence, involved in systems design work. They assessed what could be formalized in these studies of the defense system and, equally, what could not be formalized but had to be the subject of judgment both in the course of the game (simulation) and in interpreting the results of the study.[13]

The experts helped distinguish between what can be formalized and what must be the subject of qualitative judgment. This was a vital part of systems design work. As a result, the forms of work followed in the systems design process were completely different from those used in the first case. One question this raises is whether this kind of computer support can be used by members of an occupational group other than the experts whose views formed part of the systems design work.[14]

In a follow-up study on the work described above, I reflected on the experience gained. The central issue was certainty in work, that is to say the extent to which computerization improves the basis upon which decisions are made. What interested me most of all was the model's ability to represent the 'reality' with which professionals in that field were familiar.[15] In the course of system design work it became very clear that there was no apparent way of representing reality in a mathematical model. Some tension arose in the meetings of the working group, a line being drawn, broadly speaking, between the professionals in charge of the project, on the one hand, and the operations analysts with their mathematical background on the other. This tension was symptomatic of the fundamental difficulty of reconciling the inconsistencies between different specialist languages, which were, au fond, an expression of different perceptions of reality.[16] This epistemological question is a central theme of this study.

At that time – the end of the 1960s – my work was largely concerned with finding technical solutions to the problem of establishing certainty in work. The technology available at the time (batch runs on a computer) placed severe limitations on our opportunities to 'strike while the brain was hot.'[17] However, the potential of the new technology which was to come onto the market at the end of the 1960s was considered so revolutionary that a Japanese futurological study introduced a new term, the information society, to describe it.[18]

The Introduction of the 'Dialogue Computer'

John K. Kemeny, a mathematician at Dartmouth College in America, identified a number of problems in teaching computer technology; specifically, the lack of immediate contact with the computer, the considerable amount of time and money needed to learn the machine's 'language and rules' and the limited amount of computer time available to each student.[19] He found the solution to these problems, which were caused by the limitations of batch processing on computers, by developing time sharing, which was marketed by General Electric in the early 1970s as 'the dialogue computer'. John Kemeny made the following comment on the relationship between man and computer:

One can't help reaching the conclusion that it is more efficient to use a human being as the computer's partner than to spend many years trying to teach a computer a talent for which it is not well suited.[20]

The potential uses of the dialogue computer began to be discussed at a series of conferences held all over the world in the early 1970s. Typical of these was a series of symposia held at Johns Hopkins University, USA, in 1969–70 to discuss future possibilities opened up by the new technology.[21] John G. Kemeny suggested important new applications for computer technology in the public sector, for example computerizing the local employment register, the prison and probation services and the social insurance service.[22] When it was marketed, this technology was presented as a new 'tool' for professional groups such as company management, engineers, economists, production planners, administrators, physicians and researchers.[23] But there was obvious disagreement about the choice of new areas of use. Here is one criticism of computerizing the employment register, for example:

Kemeny suggests computerizing the local employment register as a good example of what might be done. It is a very different thing, however, to provide a person a seat on an airplane and to find him a job, which will match his skills and make him socially comfortable. The problems are by no means comparable.[24]

One of the purposes of this study is to demonstrate the risk of drawing false analogies between, for example, man and computers. Specialized occupational skills are developed by learning to see the differences in the manifold nature of reality, and this cannot be reduced to the brain's ability to process data.[25]

Herbert Simon, Professor of Information Processing and Psychology at Carnegie-Mellon University, presented his vision of the trends we may see in the next few decades. As a background to his description of Utopia, Herbert Simon reminds us of the inconsistency in the meaning of terms to be found in a 'common culture' and in a 'scientific culture'.[26] Introducing terms from a 'scientific culture' into a 'superficially-common language' causes considerable translation problems.[27]

The research in the field of cognitive psychology from which Simon quotes has managed to 'duplicate to a remarkable degree the human method of information processing called thinking.'[28] When a psychologist says 'machines think', the meaning he wishes to convey is determined by this new orientation of research in the field of cognitive psychology. His meaning is precisely defined and there is no satisfactory translation into ordinary language. 'If you wish to converse with him (which you well may not!) you will have to follow him into the scientific culture,' says Simon and goes on:

Make no mistake about the significance of this change in language. It is a change in thought and concepts. It is a change of the most fundamental kind in man's thinking about his own processes – about himself.[29]

Herbert Simon would like to see the highest priority given to what he calls research of 'the inner space'. He maintains there has been less research into human thinking than into planetary space and he concludes his lecture by making the following assessment:

I think we soon shall have an understanding of both the information processes we call computers and those we call man. This understanding will enable us to build organizations far more effectively in the future than has ever been possible before.[30]

Herbert Simon discusses another aspect of this question, namely the possibility of avoiding undesirable effects introducing new technology into society.

The dream of thinking everything out before we act, of making certain we have all the facts and know all the consequences is a sick Hamlet's dream,

he says and recommends using a method based on examples and case studies to give a clearer picture of the undesirable consequences of the use of technology:

The world outside is itself the greatest storehouse of knowledge . . . Of course it is costly to learn from experience; but it is also costly, and frequently much less reliable, to try through research and analysis to anticipate experience . . . The armchair is no more effective a scientific instrument for understanding this new technology than it was for earlier technologies.[31]

Using Time-sharing in Planning

At that time, in the early 1970s, I was working on the development of a series of seminars on the theme 'Using Time-Sharing in Planning'. In contrast to discussions on the possibilities of the information society which were being held in the USA and Japan at the time, these seminars were based on examples and case studies. Thus in this seminar, the business economist Åke Sandberg stated his view on decision-making in organizations:

In early management theory, decision making was regarded as a fully rational process in which formalized models could fully replace the decision-maker. Decision making in organizations is currently seen as a social process with some formalized analysis to assist the decision-maker, a view which has close links with recent developments in computer technology. This view also has links with the use of formalized analysis at higher and higher levels of decision-making.[32]

From the seminar *Using Time-Sharing in Planning*, March, 1972

In the course of these seminars I came to view time-sharing as a means of opening up opportunities for a completely new kind of planning; I saw the rapid feedback of results that was a feature of time-sharing as a crucial change for computer-aided decision making. The terms I used at the time are taken from cybernetics – the kind of terms that were used to market the dialogue computer. Here are some thoughts (which I subsequently had reason to reconsider) taken from an introduction to computerized estimate planning:

Before we can give the expression 'computer-aided planning' any content in an operational context we must define two terms: adaption and feedback. The aim is to use these terms to present a new starting point for the planning process. Time-sharing, which is the basis of a radical break with the more traditional type of planning, is an important component of this new approach.

Adaption may be defined as 'the modification of an organization (or of its functions) to make it more viable under the conditions in which it operates.' This definition is a description of the task of adaption, and not the design of its mechanism. If we want to know more about the way adaption works, we must compare the way something is done for the first time with the way it is done after it has been done many times. The first thing one notices is that in time people almost always learn to perform the action better and better. This level of adaption is acquired from case to case.

There are, of course, a number of problems concerning the practical aspects of feedback. These include the question of the form feedback should take and when it should occur. A survey on the importance of feedback of the results of a learning process, with particular regard to the optimum time after the learning process that feedback could be given, found that the more time that elapses before feedback, the worse the results of the learning process measured as progress observed in a subsequent test.

The time-sharing technique allows for rapid feedback and the opportunity to get an idea of the effects of different alternatives, even when time is very limited. The methods for increasing certainty in a planning process are closely linked with the feedback built into the routines.[33]

My thinking at the time, as illustrated in the above quote, was that 'dialogues' between man and machine resulted in a greater degree of adaption, guaranteed by the speed of feedback that time sharing made possible. I used a term from cybernetics here to develop the concept of certainty which, as mentioned earlier, my reflections on the experience I had gained from the operations analysis studies identified as being the central issue.

Computers and Skills – The Automation of Complex Calculations

In one of the case studies presented in the seminars on time-sharing in planning work, a systems engineer gave the following description of his work in the heating, water and sanitation division of a construction company. In his view, a great deal of this work could be rationalized and simplified by computerizing the advanced and complex calculations performed in this field. This would, in his opinion, make it possible to dispense with the services of design technicians and make the organization better suited to the possibilities opened up by the new technology. His thinking is an unusually vivid example of a technology-centred vision of the future.

A survey of the heating, water and sanitation department showed that 40 to 60 per cent of the work (the amount varied with the source of the information) consisted of advanced calculations and the rest of work on technical drawings and specifications.

Thanks to the close contact we had with a technician in this field who was interested in computers we were able to write software which, on the basis of the simplest imaginable input data, could perform all the advanced calculations required and could retrieve from a disk the most suitable installations in the form of radiators, pipe dimensions, etc. The calculations of quantities were, of course, carried out at the same time. In short, once these calculations were completed, the only thing the heating design engineers had to do was produce technical drawings. We felt that the architect could learn to draw in a few pipes on a drawing and copy off the dimensions and type of radiator from a computer print-out. Thus the heating, water and sanitation engineering department could be reduced to a core of staff whose job would be to update our disk-stored data with suitable registers and to keep abreast of technical advances in their field. It should also be possible to streamline other technical departments in the same way, so that what a design office would remain staffed, with more nonspecialized draughtsmen than specialist staff and a research department staffed with highly-qualified specialists from different fields, reasonably familiar with computer technology. This would make the organization more project-oriented and less divided into specialist areas, for which a project leader, let us say an architect, would be responsible throughout the design stage. He would have at his disposal a number of draughtsmen who, in addition to being able to produce all the necessary drawings, would also have a computer terminal which they could use to store the necessary data and obtain the information they require to produce their drawings.

It would be the research department's job to produce suitable installation equipment for loading the disk stations and to keep the groups of architects informed of the latest advances. Estimates and rent calculations could be carried out automatically on the basis of the information supplied by the architect's office. In the same way, the architects' office could issue orders for fittings, which would be processed by the computer and sent from the terminals to the suppliers in the required format.

It goes without saying that this requires major reorganization which, understandably, did not meet with the approval of the head of our heating and ventilation department.[34]

The systems engineer's view is based on the idea that the operator needs no professional skill to carry out the complicated heating and ventilation calculations programmed into the computer. He considers that a computer system would make it possible to transfer the work of design technicians to draughtsmen and architects.

At the seminar where this case study was presented, a systems engineer from an industrial company made the following comment:

We systems engineers go into organizations and are so terribly rational. We believe that we can make changes. I myself, with my technical background and knowledge of the company, was drafted into a computer department where I was given considerable influence on the development of routines in a certain area. We were quite confident that we were putting the correct routines in place. And that was true from the technical viewpoint; theoretical solutions are not hard to arrive at. But because we did not take into account the way people behave, our system had serious negative consequences. It did not work for these people. And that is when I discovered how little I understood myself.[35]

Let us return to the example of the heating, water and sanitation design engineers. The systems expert found it difficult to produce a reliable specification of requirements, and thought the designers should reconsider their task. He felt he had given clear definitions of certain selected terms, and if the designers had a few days' training in systems technology it would be easier for them to articulate their professional knowledge:

Years of experience have developed and modified the working methods applied in the field of heating, water and sanitation engineering to suit the capacity of the human brain, and these methods are, moreover, strongly individualized. This means that the systems engineer has to listen to the views of a number of design engineers and then try to form an opinion of the methods to be applied when the superior memory and calculating capacity of computers become available.

Now there is a phase in this process which is both intensive and a potential conflict area. This is the phase which is intended to convince experienced design engineers that they have to reconsider their methodology; to coordinate it with and adapt it to the new aids which are to be made available to them. At the same time the systems engineer is forced to question the design engineers' knowledge of what lies behind the computer systems they use.

With the benefit of hindsight, I can now say that it would have been well worth investing in a few days' training in systems technology for the heating, water and sanitation design engineers who would be involved in data collection – not to turn them into systems engineers but solely to give them a better idea of the relative importance of various aspects of the systems engineers' work.[36]

The systems engineer tried to tap the design engineers of their information. He put the design engineers' inability to describe their work processes down to an inadequate knowledge of their field. He talked of them using 'vague terms' and 'admitting their ignorance'

In my efforts to chart the working methods used, I was able to observe that many design engineers wishing to appear to be professionally knowledgeable did not even know how an elementary single pipe system worked. However, professional pride prevented these gentlemen from admitting their ignorance. Instead, as soon as I tried to elicit a response, they began to make demagogic statements on the relativity of everything. I have often met this phenomenon when professional pride is at risk, and I would therefore recommend that all systems engineers give some thought to the approach to adopt on such occasions. Should I continue to press my victim for information which he probably does not possess, or should I accept his vaguely-formulated responses and try to find the answers to my questions somewhere else? Or should I force him to admit his ignorance and then ask him to find the correct answers?[37]

A design engineer with 35 years' experience says that if the computer system is static it will become a blind alley. The system must be a dynamic one, capable of dealing with new ways of solving problems. At the same time he emphasizes the importance of concentrating mainly on a discussion of the computation principles built into the software.

Technology with a capital T emphasizes logic. It may run counter to everyday technology, which emphasizes degrees of freedom. The link between everyday technology and technology with a capital T must be dynamic, not static. This is a crucial factor.[38]

In the construction company, the systems engineer had no professional knowledge in the field of heating and ventilation. He was a technician whose task was to implement administrative rationalization measures. He was working on the basis of parallels drawn between the information processing and calculating capacity of the brain and that of a computer. His evident ignorance of the professional knowledge possessed by design engineers forced a confrontation which had serious consequences for the company as a whole. This example shows us a clash between two quite different perceptions of reality and of language. Whether these differences can be bridged, for example, by standardizing the language used is an epistemological question analogous to what I identified as being a fundamental difficulty in operations analysis study – the problem of translation between different professional or specialist languages.

Understanding a Computer System

A case study of the computerization of forest valuation work described in the next chapter was presented to the HSB (The National Federation of Tenants' Associations) at a meeting attended by representatives from both the National

Board of Agriculture and the HSB National Federation.[39] The intention was to compare the different experiences gained from developing computer systems for heating and ventilation design engineers. The interview reproduced below took the form of a discussion of the importance of understanding a computer system which will make fairly fundamental changes in the working practices. The disagreements between the actors involved stem from different interpretations of what understanding of a computer system means.[40]

The people involved were:

- *Ulf*: a department head at the National Board of Agriculture.
- *Torsten*: a systems engineer at the HSB National Federation
- *Lena*: a heating and ventilation designer at the HSB National Federation
- *Per*: a systems expert at the National Board of Agriculture.

Ulf: I would like to get back to what you just said about the changes in our understanding. I think that is a very important point. You cannot change things without understanding them. We have never brought pressure to bear from higher in the organization. I believe it is important that you understand how the whole thing works when you are designing systems like this one.

Torsten: We tried to solve this problem by providing explanations of how the forms should be completed. The file for these forms contains a definition of the terms used. If you look at a reference number you can go straight to a file and get the information you need. It is hard to imagine that things can be made any clearer than that. And yet it is difficult to get this system off the ground. Things are not running smoothly and I do not actually understand why things are so difficult.

Lena: We are looking for a different kind of understanding. It is easy to understand the relationship you describe. The understanding I assume you are talking about here is an understanding of why there should be any change at all. This is a more profound kind of understanding and I suppose that is what you are referring to.

Ulf: Where understanding is concerned, we have an understanding of the question of why one should begin to use the system, what the benefits are, how it improves quality and cuts costs, how you become more competitive, how it improves your job. This is one side of understanding. These are the reasons for building a system. The other side, the other level in this question of understanding, is an understanding of the sequence of calculations in the system. What functions do they perform, what formulae do they use, what happens in the actual computing process? These factors were quite obvious to our forest rangers. They had been doing these calculations by hand, with some help from the office staff. But they knew how to make the calculations. They knew what was going on. It was a well-developed, tried and tested system. You have not systematically gone through the evaluation methods before embarking on your systems development phase. That is one part of the question of understanding. Finally, we have the purely practical understanding of codes and the like which you mentioned here. The instructions and description of the system contains a section on the formal input requirements: what are the parameters, what code figures may occur and so on. That is what you have on that page in the file. But that is what I would like to call the level

The systems expert (photograph Peter Gullers)

of understanding which is at the bottom of this hierarchy of levels of understanding, and it is a superficial understanding of the system. The essence of the system, or what the system aims to do, so to speak, is what the deeper understanding is about, and that is what is important whether or not the system saves time or money. I believe the point of the system is to save money in the properties where your products are going to be used. Heating costs will be cut, and that is particularly important in the present energy situation. A very important goal is to produce quality products, products which live up to our.

Per: Yes, in one respect our users do not have a complete understanding of the system, inasmuch as they did not understand why they were supposed to sit at the terminals, and so they did not do so. Then a new category of people came into the picture: the operators. They were almost forced into this job and in the beginning we did not do very much to help them understand it either, and that is what the criticism of our system has been – 25 operators were not given the training they needed to have an underlying, more profound understanding of why these operations have to be carried out. They must have been given a superficial understanding so that they were able to press the right buttons. Now, after being criticized on this point and having realized that the situation is not good, we are building up this in-depth understanding. We can say that there are people who run evaluation programmes without having any deeper understanding of them. What do the results mean? Our evaluators did not understand – and still do not understand – the point of sitting down and using the computer to carry out their calculations.

Lena: I had a question earlier, about the operators versus the evaluators. It was originally intended that the evaluators would use the terminals; is it true that none of them do so?

Per: Well, there are a few exceptions.

Lena: Why did they not want to use the computer?

Per: It is quite a complicated issue. One way of looking at it is to say it is because they cannot type. These are practical people who work in the field. But I do not believe it is a simple as that. I believe that there are far deeper reasons. Perhaps they feel that sitting at a keyboard is beneath them. It is obvious that some of them feel it is degrading. They do not want to do office work.

Per, the systems expert at the National Board of Agriculture, sees the reason for this 'keyboard rejection' as an indication of a problem embedded in the working culture of forest rangers. The difficulty does not lie in the act of sitting at the computer and pressing keys.

In the next chapter we follow the systems expert at the National Board of Agriculture and examine more closely the effects of a computer system on a professional practice.

Notes

1. KLINE, Morris *Matematiken i den västerländska kulturen*, Prisma, 1968, p. 152.
2. HYMA, Anthony *Charles Babbage: Pioneer of the Computer*, Oxford University Press, p. 256.
3. For an account of developments in the later stages of this project, involving the Swedish lawyer and book printer, Georg Scheutz and his son Edvard, see a study by Michael Lindgren: *Dator för 150 år sedan – En historia om ett misslyckande*, in Bosse Sundin (ed.): *I Teknikens Backspegel. Antologi i teknikhistoria* (In the Rearview Mirror of Technology: An Anthology of the History of Technology), Carlsson Bokförlag, 1987, pp. 233ff. See also Michael Lindgren's doctoral thesis: *Glory and Failure*, Linköping Studies in Arts and Science, 1987.
4. *Ibid.*, pp. 244ff.
5. See BOLTER, David *Turing's Man. Western Culture in the Computer Age*, Duckworth, 1984, p. 34.
6. FOA Operationsanalys. FOA orienterar OM, No. 8, 1967, pp. 13ff.
7. GÖRANZON, Bo *Tabeller över lösningar till Lanchestersekvationen*, FOA P-rapport, C 8164–2, 1967.
8. FOA, 1967, p. 13.
9. Lanchester's differential equations are a mathematical model of the so-called analytical type. They express the relationship between the input data and the results directly in mathematical formulae.
10. GÖRANZON, Bo *Administrativt systemarbete*, Armeéstabens OA-rapport, C2, 1969.
11. *Ibid.* pp. 31–38.
12. The mathematical formalism was extensive and contained a complex pattern of calculations. As the method was developed, the calculations were set down in a notation which could be transferred to a computer. This work was carried out over a two-year period.
13. The mathematical planning of the methodology made possible a division into clearly separated operations, the content and function of which were uniformly defined. It was natural to divide the calculation pattern into so-called sub-routines in the FORTRAN programming language. The complex calculation pattern involved about 10,000 assignment statements and a volume of indata of about 20,000 data units.
14. The computer support for decision-making introduced in this example was used only once by the people involved in the development work. See also the intervention of TOBIN, N.R. and

BUTFIELD, T.E., OR Branch, British European Airways, IFORS Conference, August, 1972 in *Svenska Operationsanalysföreningen. Proceedings*, Tobin and Butfield emphasized the importance of tailor-made decision support systems.

15. GÖRANZON 1969, p. 16.
16. *Ibid.* p. 38.
17. See HEDBORG, Bo *On man-computer interaction, an organizational behavioural approach*, BAS 1970:6, Göteborg and HEDBORG, Bo *Dialogdatorn – sjuttiotalets räknesticka?* (The dialogue computer – the slide rule of the seventies?) in *Proceedings of the STF Engineers Training Course,* 1973.
18. Japan Computer Usage Development Institute: *The Plan for the Information Society – a National Goal Toward Year 2000*, Computerization Committee, Final Report, May 1972.
19. HARTZELL, Svante *Honeywell Bull Time-Sharing Service – a broad presentation* (information brochure) 1971, p. 1. General Electric developed a communications system which could simultaneously serve 39 users of slave terminals. This system allowed for close contact with the machine as the students were now able to give direct orders to the computer, and simultaneous use gave more computer time per pupil. The BASIC (Beginners All-purpose Symbolic Instruction Code) programming language, with 17 instructions, was developed for this system.

    General Electric recognized that there was a new market for user-friendly, uncomplicated 'computer power', and this innovation spread rapidly. By 1965 two systems were available for public use in the USA. In 1968 Sweden had its first Time-Sharing system at the same time as it was introduced in ten countries in Europe, Japan, Australia and North and South America. At the end of 1971 a further development was introduced, where computer power was supplied via satellite from a large computer complex in the USA. It became possible to use national and international information systems which were of considerable interest to organizations spread across the globe.
20. The quote from Kemeny is taken from Bolter, 1984, p. 234.
21. A series of symposia, held in 1969 and 1970 at Johns Hopkins University, USA, discussed future perspectives opened up by new technology. These discussions are to be found in Martin Greenberger (editor) *Computers, Communications, and the Public Interest* The Johns Hopkins Press, 1971, p. xv.
22. *Ibid.* p. 20.
23. *Ibid.* p. 39.
24. *Ibid.* p. 40.
25. *Ibid.* p. 40.
26. *Ibid.* p. 52.
27. *Ibid.* p. 52.
28. *Ibid.* p. 47.
29. *Ibid.* p. 129.
30. *Ibid.* p. 130.
31. *Ibid.* p. 131.
32. SANDBERG, Åke: *Perspektiv på organisationers beslutsfattande* published in GÖRANZON, Bo (Ed.) *Planeringsarbete med Time-Sharing*, Honeywell Bull, September 1972.
33. GÖRANZON, Bo: *Modellutveckling* published in Göranzon, 1972. A reflection on this work is described in GUILLET de MONTHOUX, Pierre and GÖRANZON, Bo *Beskrivningsaxlar för informationssystem*, Ind EK ORg, KTH, 1973a (*arbetsrapport*).
34. This quote is from GÖRANZON, Bo (Ed.) *Datautvecklingens Filosofi. Tyst kunskap och ny teknik*, Carlssons 1983, p. 46.
35. A tape recording of a seminar on Planning Work with Time-Sharing, September 1972, GUILLET de MONTHOUX and GÖRANZON, 1973a, p.42.
36. *Ibid.* p. 48.
37. *Ibid.* p. 48.
38. *Ibid.* p. 48.
39. GÖRANZON, Bo *Studier av Arbetsorganisation och Datasystemutveckling Kunskapsuppbyggnad vid systemutveckling genom en jämförande analys av två fallstudier, PAAS arbetsrapport* No. 4, Ind Ek Org, KTH, Februar 7.
40. *Ibid.* pp. 65–83.

Chapter 2

The Computerization of a Work Process: A Case Study of the Long-term Effects on Professional Skills

The Research Method

The transcript of the taped discussion at the end of Chapter 1 reveals an unforeseen consequence of the computerization of the County Agricultural Boards. It was the office staff, and not the forest rangers, who manned the terminals.

In this chapter we examine why the forest rangers were unwilling to use computers in their work. It is a story of contrasting perspectives and contention, both at the time the computer system was introduced and when it was updated and expanded some years later. It is the story of how computerization affected the professional skills of the forest rangers and the office staff, which were the two occupational groups involved. This case study – begun in 1974 when the computer system, which was one of the first applications of time-sharing in Sweden, had been operational for barely five years – had a decisive effect on the studies of the long-term effects on skills I carried out from 1975 to 1989. The discussion reported in Chapter 1 gave some indication of the research method developed and used in the course of this case study.

The philosopher, Tore Nordenstam, made a vital contribution to the development of the theory and methodology used in the study. Nordenstam's study, 'Sudanese Ethics' had a direct influence on the method we used.[1]

When I was working on articulating and analyzing Sudanese moral attitudes, I taped every interview with the key people (informants), transcribed the tape as soon as possible after the interview, reflected on the transcript and then used it as a starting point for the next conversation. When working directly with informants one can have one's impressions constantly confirmed or corrected by the people involved.[2]

The main point in this method is the reflection on what the interviews and discussions have produced. This is an ongoing process throughout the descriptive and analytical phases. Once the work has been completed and the report produced, there is further opportunity for confirmation and refutation.

The people who are directly involved can express their views on the findings of the survey as a whole and point out gaps or distortions in the report, produce examples to the contrary or add further examples which coincide with the suggested pattern. Other people can also become involved in this process. Feedback on the research findings, descriptions and analyses passed on to those who are the subjects of study is not just a matter of moral concern. It is an important part of the hermeneutic method itself.[3]

The picture becomes somewhat more complicated when one is dealing with several information sources, but the basic principles are the same as those in the simple interview situation described in 'Sudanese Ethics', when there were no more than three people providing information. The preliminary work included a number of seminars and interviews with the leaders of the National Board of Agriculture, the forest officer in charge of systems, representatives of the forest rangers and the assistant staff at the County Agricultural Boards and representatives of the trade unions involved. These seminars and interviews were carried out in 1974 and 1975 and gave dissimilar pictures of the sequence of events. Some of these interviews were taped. A number of questions gradually emerged and were used later in the study.

Aids such as standardized interviews, occupational group meetings and recorded discussions were used in order to maintain continuity in the interviews with so many informers.[4] All the forest rangers and office staff in the organization were interviewed in 1976.[5] Another major influence on the method used in the case studies was Albert Danielsson's study of business administration.[6] Danielsson demonstrated how the real-life situation of a business does not make up a logical whole. This also holds true for business administration. Instead, departments, routines and methods emerge to parry external threats to the company. These responses to urgent problems have then become a part of the company, in the form of what Danielsson called a kind of sediment. This sedimentation principle is in stark contrast to the accepted principle of rationalization.[7] A researcher who accepts the sedimentation principle must, as a consequence, begin to dig into the history of the business he is examining. This may provide the basis on which to build a complex picture of a chain of events in contrast to the rationalization principle, which has to arrive at uniform conclusions.

The County Agricultural Boards and Forest Valuation

The executive bodies under the National Board of Agriculture are the County Agricultural Boards, of which there is one in each county. Their main task is to work on the rationalization of agriculture and forestry, i.e. to create rational (profitable) agricultural and forestry businesses. This work includes the purchase and sale of agricultural and forest land and properties. About 1000 purchases are made a year, totalling 40–50,000 hectares (1 hectare = ca. 2.5 acres). The land purchased is sold to farmers whose businesses would benefit from expansion.[8]

The County Agricultural Boards estimate the value of the property in question for these transactions. The work is carried out in the following stages:

The value of the property is estimated on the basis of a description of the forest, produced by trained staff (forest rangers) who carry out an inventory in the field. The inventory is made by dividing the land into sections which are as uniform as possible. These sections are marked out on a map, and then described in terms of potential production, the size and composition of the stand of timber and the possibility of felling.

A forest ranger using a stereoscope. Before the field evaluation, the valuer examines aerial pictures of the forest through a mirror stereoscope to get a good picture of the proportions of cleared ground, young trees and mature forest. Preliminary boundaries for stands of forest are sketched in on the aerial photographs.

The next step is to determine the area of the sections with the help of the map and to register this information on a computer medium. This information is keyed in to computer terminals. Lastly, the terminals are connected to the host computer and the value is calculated.

In the forest. Data gathering does not always take place under the best conditions. At a temperature of −20°C, and with a recent fall of snow, an assessment must be made of whether the forest may be considered to be adequately developed or whether there must be costly additional planting. The parcel of forest is valued. The valuer makes an inventory by mapping the forest and assessing its age, volume, cost, position, etc. The assessments are entered as data on special forms designed for data processing.

A forest ranger doing a manual calculation. This process requires a great deal of know-how and time. But it has the advantage of giving the ranger first-hand knowledge of the property.

At the terminal. The field inventory forms are given to the assistants for keying-in and transmitting through the telephone network to the host computer in Stockholm.

At the farm. Using the valuation as a basis for negotiation, the transaction is discussed with the owner. The formal, detailed computer printout often commands unwarranted respect.

The Main Sources of Information

In the course of the events which will be discussed here, and which led to the situation we have today, opposing standpoints, disagreement and conflicts emerged on central issues such as the quality of forest valuations and the issue of who should be held responsible for a valuation when part of the work is done on a computer. Here there were different basic perceptions of what constituted a judgment and the ability to make judgments.

The people presented below were the principal sources of information on the chain of events reported here. To some extent they also represent different phases of the development of the situation. In the first phase the County Agricultural Boards were set up and a method established for forest valuation. The second phase was the development and introduction of a computer system into what was by that time an established valuation procedure, and in the third phase a new forest valuation method was introduced, and developed on a completely different basis to the former method.

Lennart Kallstenius: Head of the Forestry Department at the Central Office of the National Board of Agriculture and responsible for setting up the County Agricultural Boards in the 1950s. His job included developing a method for forest valuation. He was influential in drawing attention to the long-term effects of computerization on the professional skills of forest rangers. Lennart Kallstenius retired in 1975.

Ulf Larsson: Head of the Forestry Department at the National Board of Agriculture Central Office since 1975.

Per Svensson: A forest ranger on the Forestry Department staff with the job of developing and introducing a computer system for forest valuation. Per Svensson brought up the issue of professional ethics and responsibility in the context of the computerization of forest valuation work.

Per-Johan Åge: A forest ranger on the staff of the Forestry Department with the job of developing forest valuation methods. He described the allocation of work between the forest rangers and the office staff underlying the development of a method for forest valuation.

Gun-Marie Forsberg: A cartographer and calculation assistant at the Umeå County Agricultural Board, who became a computer operator. She formulated the computer operators' criticisms of the employer and the trade union, notably the fact that there was no training given and no appraisal of the new, more complex work tasks when the computer system was introduced.

We have already given a brief description of the current computerized property valuation procedure. The basic conditions for this work, which began in 1968, were summarized in 1975 by Per Svensson, the systems expert, as follows:[9]

(i)   The forest valuations carried out by the County Agricultural Boards and which, among other things, are used as the basis for buying, selling and exchanges, involve a considerable amount of calculation, most of which is purely routine in nature. At certain points in these calculations the valuator must use his knowledge of the particular characteristics of the property to make plausible judgements and make corrections to the calculation on the basis of the interim results, continue thereafter with the main calculation.

(ii) The amount of data to be processed is relatively small, but on the other hand, the calculations to be carried out are extensive and complicated. The sequence of calculations comprises close to one thousand operations, which means that a manual calculation of the value of a property takes seven to eight hours. About two thousand such valuations are carried out each year.

(iii) It must be possible to interrupt the computation to allow the valuator to make an assessment of the compilations presented as interim results. Using these interim results, the forest officer must either order the process to continue without change, or introduce corrections and then let the calculation work continue. The judgments which must be made in this way are important because the officer has no total picture of the property at the field inventory stage. The total picture is a synthesis which emerges in the course of the calculation.

(iv) When the final result is presented, the valuator should have the opportunity to carry out alternative calculations. A valuation is a preliminary calculation based on numerous more or less uncertain assumptions. A more reliable valuation of the property is obtained if the valuator has access to several computations of the value, each made using different basic conditions.

(v)   It must be possible to present the results of a forest valuation very quickly, in extreme cases on the same day as a field inventory is completed.

(vi) Because the results of forest valuation calculations need to be presented very quickly and the calculations need to be interrupted for some recalculation. This means that the traditional process cannot be used. A computerized system was not developed until completely new technology had become available. The first time-sharing system was introduced in Sweden in the autumn of 1968. The National Board of Agriculture decided to carry out trials in producing its own programmes. The programming work began in June 1969 and trials were run in December of the same year. Full-scale production was begun at a number of County Agricultural Boards in March 1970.

The points put forward above define how the systems expert sees the forest rangers' role. Among other things, he mentions plausibility judgements which must be made in the course of the computation, irrespective of whether they are carried out manually or automatically. Unless this judgment is made when the computation is partly completed there is no point in proceeding.

Per Svensson also says that the calculation itself gives the forest ranger an overview of the property that he does not get from the field inventory, and this is an important factor for the rangers. He also says that the calculation is not something absolute. Making several calculations with varying input values to reflect variations to the original conditions gives the forest ranger a better basis on which to make a valuation of the property. Per Svensson's attitude reflects the tradition to which he belongs, namely, that the computer should do the calculating, but that the forest rangers should continue to make the judgments.

The development of a Forest Valuation Method

The first modern agricultural policy programme was produced in 1948, which was when the County Agricultural Boards were set up. At that time there was no accepted way of putting a value on afforested land. Lennart Kallstenius, the newly-appointed head of the Forestry Department, had to tackle this issue and develop a system for valuation. The following quote are his comments on these pioneer days:

When I began to examine this area more closely it became apparent that nothing had been done in terms of forest valuation. There was a general lack of standards. Something had to be done, and the great advantage we had was that we could create something from our own ideas without being encumbered by a lot of dogma. In other words, we could get started fairly quickly. We could get our proposals approved and then build up a body of experience.[10]

I began by making a tour of the country's twenty-four counties, interviewing the County Agricultural Boards and some of the big forest companies to find out about the problems in this area. On the basis of my notes I then outlined a method for dealing with valuation matters. This outline included a valuation system.[11]

The term 'valuation system' refers to a mathematical method of calculating the value of afforested land. It took a long time to develop a method for forest valuation: nearly fifteen years passed before it took shape and was documented under the name of 'The Plan Method' in 1967.

As the network of County Agricultural Boards was built up, forest rangers were employed throughout the country. At first they had no written instructions and their introduction to the work was in the form of informal conversations; a kind of apprenticeship training. Lennart Kallstenius described this process:

We went out into the forest together, taking with us stencilled forms giving us certain guidelines. I did a lot of travelling at this time. I had a very intensive dialogue with these officers. I had to work entirely on my own for the first year. I travelled round the provinces and out into the forests and made valuations. In other words, I had practical, first-hand contact with the problems.[12]

In the expansion phase, the forest rangers were largely recruited on the basis of personal assessment. Lennart Kallstenius compares the way rangers were employed in his day with the formal system which replaced it:

It was obviously undemocratic by today's standards of staff recruitment. I quite simply hand-picked people through the contacts I had with the forestry colleges. I discussed with them who would be suitable for this job, which was not primarily an administrative one. Only about ten per cent of the people who graduate from forestry college have an aptitude for this kind of speculative business activity. In my view, people who had some link with forestry had a definite advantage. I felt they would have a better understanding of general public reaction. There was some opposition to this recruitment principle. Staff organizations have increased their influence in recent years and have made it more difficult to hand pick people in this way.[13]

These two comments tell us a great deal about the way the method for forest valuation was developed. The following are key phrases: 'practical, first-hand contact with the problems', 'speculative business activity', 'a better understanding of general public reaction'.

Lennart Kallstenius insisted that a method had to be developed which was easy for the general public to understand. During this phase, as the method took shape, relations with the general public were good. The open attitude of the forest rangers had a positive effect and as the method was developed, Kallstenius and the forest rangers maintained a continuous dialogue with the public. The technical terms that came into use during this phase helped make this communication more effective.

The basic attitude during this lengthy development phase was that forest valuation was not a question of which calculation technique to use. Kallstenius was working on the development of methods of calculation, but for a number of reasons he felt that one should not be too tied down by them. It was not enough to work out the projected growth and yield of a stand of timber, but there was also the question of selling the property.

When calculating yield, it was just as important to use the opportunity to produce sales value statistics, i.e. being able to see the real situation – the way people bought and sold areas of forest – and to try to establish what variations there were from one property to the next and how they affected the purchase price. Is there a link between different kinds of business and different purchase prices?[14]

Kallstenius saw a difference between calculation and judgment:

Valuations are not just a question of making calculations, but also a question of making judgements with the guidance of the calculations.[15]

This statement is a key to Kallstenius's attitude to the formalized mathematical method as an aid in this work.

Standardization

Before the computerization of the arithmetical calculations in forest valuation at the end of the 1960s, some changes had been introduced at the County Agricultural Boards. One change was in the allocation of work – the office staff helped the forest rangers with the calculations – and another was a sharp increase in the volume of work.

Yet a third and important change, was a directive issued at the national level requiring a certain amount of standardization in the way in which the method was applied.

In the pioneer period, when there were virtually no established routines, all phases of work were carried out by forest rangers. But from the mid-1950s, when the routines for making the calculations became more stabilized, the office staff began to take over more and more of this work. By the mid-1960s, when a complete set of valuation instructions had been produced and a series of printed forms introduced to standardize calculations, some County Agricultural Boards had transferred almost all the mathematical work, including the more advanced calculations, to the office staff.

The demand for standardization came as a response to the development of the method. This comment by Lennart Kallstenius summarized one view at the time:

Forest rangers may be individualists who work under different conditions, but we have to standardize our work to some extent.[16]

He also said that the staff at the Central Office should control the way the method was developed and applied.

There is a risk that forest rangers working in the County Agricultural Boards will modify the methods used to make the calculations. We do not accept that the forest rangers make experiments, we want the experimenting to be done at the national level. But the forest rangers may put forward ideas for us to test.[17]

The beginning of the 1960s saw a sharp upswing in the turnover of agricultural and afforested land, and the organization grew. The number of forest rangers increased from twenty in 1960 to fifty in 1970.

What Were the Effects on Judgment Over Time?

As mentioned above, the idea of using computers was raised by the increasingly heavy work load placed on the forest rangers. As there was a ban on recruiting new forest rangers, the only option was to rationalize their work by introducing new technology. It was assumed that computerization would render the administrative procedures more effective.[18]

We have already described the conditions obtaining for systems development work in the mid-1970s as set out by Per Svensson. As far as possible, the computer system was to reflect the procedures currently in use.

Systems development work began with this directive:

Take the written instructions, the 'plan method', and have the computer do the work the valuers and office staff do by hand. No changes and no additions: try to transfer the manual calculations to the computer.[19]

And that is what happened. With the exception of some technical modifications, the main principles of the plan method were computerized intact. However, another source, Per-Johan Åge, said that the discussions on computerization at the end of the sixties interrupted the process of refining these calculation processes.[20]

Because Per Svensson based his work on the assumption that the forest rangers would make judgments of the plausibility of the calculations before moving from one stage of the computation to the next, his programme included questions after each stage:

There were a large number of questions in the system, about thirty, on changes in the conditions on which the calculations were based. The forest rangers were supposed to work in the same way as they did when they carried out their calculations by hand.[21]

Another factor brought to the fore by computerization was a matter of professional ethics, namely who was responsible for the valuation when the calculation was carried out by computer. Per Svensson's comment is in line with the role he thought forest rangers should play:

I distanced myself from the issue of responsibility at an early stage by requiring a signature from the officer in charge of the matter. I did not accept responsibility for the valuation unless I had his signature. By signing his name, he assumed responsibility for the figure that was produced. There is no system that is completely free of errors. . . . You cannot do without individual assessments and plausibility judgements.[22]

The decision to computerize was taken in 1969. Lennart Kallstenius firmly believed that a computerized system should be subordinate to the judgments made by the forest rangers. The following quote may be interpreted in the light of his view of valuations as 'judgements made with the help of calculations':

We employed a forest officer whose full-time job was to see what how time-sharing technology could help solve our problems. . . . I made it an absolute requirement that our need for advanced computations had to be met by computer technology. Otherwise we were not interested. Far too often in our society we have to compromise on quality for financial reasons. Many processes are irreversible, and this is an important fact to bear in mind when we cross these boundaries, when considerations of quantity or precision take precedence over quality. Our society does not benefit from levelling off in all areas.[23]

Kallstenius expressed his fear of the long term risk that forest rangers would be unable to maintain their level of professional skill. He made the following observation on the importance of manual calculations:

Are we depriving our forest rangers of the opportunity to gain the experience they need? The manual calculations they do in the office teach them which factors are important and which are less significant. The cause and effect relationship throughout the sequence of operations becomes apparent to them. This is the best way of seeing clearly how some factors in the process affect the final result. This is essential experience for people who are going to make judgments out in the forest. Will this experience be lost to us when we computerize the calculation process? I think there is a risk that it will.[24]

Kallstenius reminds us of the importance of making an overall assessment, and by that he means placing the valuation in the perspective of a sale and a purchase. He emphasized the importance of keeping the facts in mind in order to see the links between cause and effect in a complicated pattern of calculations. This is more important than being accurate to one more decimal point.[25] His attitude to computerization was revealed in an observation made at a meeting with the forest rangers in 1977:

Practical and commonsense know-how could be at risk among the younger generation of valuers because this is a comfortable way of using computer technology.[26]

In his summary, Kallstenius raised an issue which is crucial to this case study: What happens over time to the ability to make judgments when calculations are computerized?

But there were people in the organization who held different opinions. Per-Johan Åge was one of those who had a different attitude to professional skill. Already at the beginning of the 1960s he was saying that the office staff could use these well-developed algorithms without having to be familiar with or understand the underlying method for forest valuation. In other words, he took the same view as the systems expert in the HSB National Federation mentioned in Chapter 1.

The Intermittent Function – A New Occupational Group

The computer system and EDP routines were installed in some of the County Agricultural Boards in 1969. But five years were to pass before they were widely used in the organization. As intended, the system was introduced at a leisurely pace and on a voluntary basis.

However, the valuers never used the system themselves. Only two of the forty-odd forest rangers who initially had access to the terminals actually learned to use the system.[27] The staff who helped the forest rangers with the calculations had to learn to use the computer system, and ran the calculations on the terminals for them. Particularly at the County Agricultural Boards, where the office staff made the more advanced computations, it was considered natural for them to take over the work of running the calculations on the computer.

This meant that the idea behind the systems development work was never realized. The forest rangers never had a 'dialogue with the computer', because they never changed the input data to produce a reliable body of basic data for their valuations. Thus, the questions the systems designer had put into the system lost all their meaning:

Then there was a problem with the questions which I had put into the programme. The computer operators couldn't answer them because they didn't have the necessary knowledge of forest valuation methods. They complained that there were a lot of questions to which they had to answer no. They wrote 'no' twenty times, as the valuers told them to do. The valuers spent the time they had gained on negotiations. . . . The problem was that they didn't want to sit at the terminals. They all can, but they don't want to. They will lose their grasp of the calculating processes if they don't know anything about what is going on.[28]

We had 25 computer operators. . . . Today our problem is to train the operators and maintain the valuers' competence levels so that the computer does not turn them into robots. They don't know what a valuation is. The operators have to be trained to do more than just a mechanical job which they know nothing about. They have to understand what it is they are doing.[29]

Thus computerization had an unseen effect on the professional skills of both the forest rangers and the office staff. The study showed that when the computer system had been in use for some years the forest rangers' professional skills showed signs of 'erosion'. The office staff, employed up to then as cartographers and office workers, had to adapt to a new professional role – that of computer operator. This group of employees would play an intermediate function between the computer system and the forest rangers, a development which no one had foreseen.

The office staff were strongly critical of introducing computers. The first time the group of twenty-five office staff working on cartography and calculations had to deal with the computer system was in autumn 1970. A systems engineer from the computer suppliers ran what was a standard course covering elements of computer technology such as describing a variable, matrices and loops, learning to write programmes and using the terminal. This course was taken from the basic school syllabus with a section added for the operators with office skills and training in cartography. Gun-Marie Forsberg had this to say about the course:

I was able to follow the first half day of the course. I understood nothing they said after that. When we asked for explanations of specialist terms, they used even more specialized terms and jargon.[30]

A few weeks after the course, the systems expert visited the operators at the National Board of Agriculture to give them information about the computer system.

He told us which buttons to press to run forest valuations. When we asked for explanations, we were virtually told that it was so complicated that we shouldn't try to understand it. 'Do as I say, and don't ask so many questions'. It was six months before I understood the difference between a file and a programme.[31]

In 1974, when the computer system had been in operation for four years, it was suggested that a revised version be introduced, the idea being to bring it on line in February 1975. Gun-Marie Forsberg made this ironic remark during a seminar in the autumn of 1974:

Instructions on punching and coding were circulated to the Department of Agriculture forest rangers well before February 1st. Only a few of the forest rangers had ever run a valuation during the five years the system had been operating. The letter accompanying the instructions to the forest rangers said: 'It would be valuable for the operators to take part in demonstrations and studies of the material which is now available.'[32]

During these five years the operators had gradually become trapped. Because they were not employed as computer operators and had no formal qualifications in data processing, they could not apply for employment as terminal operators in other companies. Some who were qualified cartographers found it difficult to obtain such employment in other companies because they had not carried out this kind of work for several years.

As things are now we don't feel that we belong anywhere. Many of the operators feel that they have been put in a position which they cannot get out of.[33]

In autumn 1975, Per-Johan Åge took part in a discussion on the introduction of the computer system.[34] He said he felt it was not possible for a government body to train people to take on more advanced work. A government body can only provide training for the work tasks their employees perform. Gun-Marie Forsberg said:

When they talk to us about training, it sounds as if they were discussing a five-year programme. That is not what it is about at all, it is a question of building up certainty so that you know what you are doing.[35]

This is a problematical viewpoint, not least because the majority of the office staff had expressed a genuine lack of certainty, caused by an incomplete understanding of the theory of forest valuation. At a meeting of the operators arranged a year later, they considered that the course should focus on extending their competence in valuation methodology. They also emphasized that was important to be familiar with both the practical and the theoretical background to forest valuation before beginning to work on data processing. Lennart Kallstenius said that the operators must serve a kind of apprenticeship with the forest rangers to learn the basics of the valuation method by practical experience. Another important demand was for the introduction of formal positions as 'computer operators'. The title of computer operator applied to themselves was thought to be misleading, and it was suggested that they should be called forestry computer technicians.[36]

A Growing Conflict

A description of the forest rangers' position after the introduction of the computer system should mention the conflict which emerged at the County Agricultural Boards. The protagonists were people on the Central Office staff, but the conflict was very much to do with national local relationships.

This conflict sharpened towards the end of 1974, when the County Agricultural Boards had had some five years' experience of using the computer system and were in the process of producing an updated version of valuation procedures. As a researcher, I was brought in to interpret and describe the growing disagreement centering on the computer system and the method of forest valuation. Lennart Kallstenius wanted to know why the forest rangers did not use the computers themselves.[37]

The underlying reason for changing the valuation process was that the County Agricultural Boards had to adapt to meet new economic circumstances in the forestry sector. A new Forestry Act had been passed which meant that some changes had to be made to the principles for surveying and valuation in order to bring them in line with new quota regulations. The valuation method was revised and new parameters were added to the model. This involved a significant departure from the original valuation method, evolved largely from experience-based knowledge and built up and tested over a long period. The plan method, as it was known, was more than a document of the algorithms for forest valuation. It was documentation of the development of the valuation process in a 'living' discussion since the early 1950s. If a broader group of people were to gain a better understanding of matters related to forest valuation, then the meaning of the terms used and the principles applied must be understood by both buyer and seller.

The conflict at the National Board of Agriculture centred on the relationship between the forest officers responsible for the new theory of forest valuation, led by Per-Johan Åge, on the one hand, and their colleague Per Svensson, the senior systems expert, on the other. The conflict was triggered by the scheduling of the introduction of the revised valuation method. Per Svensson needed time to programme in the new parameters, and the forest rangers wanted a planned course of training to allow them to learn about the changes that had been made to the valuation system.

Per-Johan Åge criticized Per Svensson for the way he had handled the systems development work. He said Svensson was in a strong position because 'he has deliberately avoided training anyone else, so he is the only person who understands the computer system.'[38] Svensson was also criticized for not producing documentation for the computer system. This was yet another reason why it was impossible to understand the systems development process.[39] The strength of Svensson's position lay in the threat that he might leave the organization.

Per Svensson's criticism of the Central Office staff was that he alone had carried the responsibility for the systems design and programming and that involuntarily, his job had become more like that of a 'computer expert'. This reflects the deeper differences that lay behind the controversy.

The other staff members showed very little interest in the technical and practical problems of the computer system. The conflict grew from different views of the way a computer system should be used, or rather, the way the people who work on a computer system should be regarded.[40]

The central staff, and Per-Johan Åge among them, held that the computer system could be used like a desk-top calculator, without any deeper discussion. Per Svensson thought, however, that the computer system was a completely new way of working which had enormous potential for both producing positive effects, and also for misuse. It was therefore essential that the systems development strategy and the use of the system in valuations should be discussed.

Automated Cost Estimates

One example of the revisions to the valuation method was an attempt to automate the valuation of cost estimates for felling.[41] For a long time the forest companies had made surveys and produced estimates for forecasting the cost of using different logging techniques – either fully mechanized or the traditional chainsaw or manual felling methods.

They had established that certain factors such as the volume/hectare, average trunk diameter, proportion of tree trunk with branches etc., has a significant effect on the time required for the felling operation, and thus on its cost. It is therefore natural to apply this experience in forest valuation, using a computer to calculate the costs on the basis of date collected in the field. This required a great deal of work to produce a summary of experience in a processable form and to write new indata routines for the necessary particulars, and having these control a number of mechanical functions which would then produce the cost of felling per cubic metre.

The input data form looked roughly like this:

Transport	Vol/ha. Cost of tree types	Prop.	Diameter branches	Prop. of Land type	distance	
A	BCD	E	FGH	I	J	K

The details A to J had to be filled in the forest so that the programme could calculate value K, the cost per cubic metre. The routines were designed to allow an assessment of cost K to be entered so that the cost for small or 'impossible' stands of forest could be assessed immediately. After the system had been tested and found to work, it was put into effect. Per Svensson observed:

But when analyzing how the system was being used a couple of years later, it was found that only one in fifty valuers filled in the data needed for the automatic calculation, while the other 49 entered the volume cost figure straight away. The system required a zero to be entered in all the columns that were not used, so the computer operators spent one week of each year keying in zeros![42]

The same thing happened with the original plan method when the 'dialogue' procedure was removed from the system. In this example, an automated cost calculation, the programme was changed and the data for automatic calculation were removed. Per Svensson said: 'the land survey model for forest valuation, the so-called 'table method' and the County Agricultural Boards' 'Plan method' are being combined to form a single model, and the automated technology will be reintroduced.'[43]

Why was the 'expert system' – automatic cost calculating – not used? Per Svensson provides some explanations:[44]

(i)   The forest rangers did not trust a calculation which was produced after they had made their survey of the forest. They wanted to set the price themselves.
(ii) The only forest ranger to use the system is one of the most experienced rangers who knew enough about the automatic system to tell from the data entered in columns A to J what the final figure would be.
(iii) No two stands of forest are alike, so standardized calculations cannot be correct.
(iv) Because some factors, the local availability of hauliers, for example, were not included in the calculation, it could not produce a correct result.
(v)   Expert systems can only be used and assessed by experts, who do not really need them. Non-experts cannot judge the results of the system and should not therefore use the system before they have become experts themselves, at which point they will no longer need the system.

The Local View

Most of the differences in attitude to the computer system concerned perceptions of the relationship between the national level staff and the forest rangers who worked at County Agricultural Boards all over the country.

As mentioned above, Per-Johan Åge and other Central Office staff stipulated that there had to be some standardization in the way the method was used.

Although the forest rangers had a strong sense of professional identity, the standardization requirement caused some contention between the national and local levels in the organization, and this gradually escalated into open conflict.

Per Svensson felt that he had to assume responsibility for service to the forest rangers, 'so that they will get the support they need to carry out their difficult tasks.' He emphasized the independence of the forest rangers, based on the fact that it is the rangers and not the central staff who are responsible to the clients:

The valuers work independently of the Central Office staff. It is their job to negotiate with the owner of the parcel of forest land and justify the value they have arrived at as being reasonable. The forest rangers are the local-level valuation experts, and the National Board of Agriculture quite simply cannot have the same experience of the local market. By and large, they decide for themselves how they will do the calculations. They must make their assessments and they must have enough confidence to sign them. It is they who are responsible. This means that in future it will be difficult to say that everyone has to use the terminal for data processing. It is used by the people who want to use it.[45]

It is worth noting that in this comment, made in 1974, Per Svensson emphasized the responsibility the forest rangers have as civil servants, that is to say vis-à-vis the general public. The forest ranger must be given considerable latitude to decide what method of calculation to use or whether or not he wants to use the computer system.

The conflict between the Central Office staff and the forest rangers centred largely on ways of dealing with local conditions which could have an effect on the forest valuation, for example the type of trees, forest growth, the local market, etc.[46] The following statement made by one of the forest rangers illustrates the nature of this conflict. In it, he rejects the idea that the factors on which the valuations are based should be used to exercise central control over forest valuations:

It seems to me that there are risks in computerizing too many of the factors on which the valuations are based, for example time formula calculations and standardized prices and then making corrections to adjust them. We have a third party to consider and we often forget that fact. Serious problems arise when we explain the sequence of calculations and the factors on which the valuation is based to a third party, who often has no training or experience in this area.[47]

For this particular forest ranger, the accepted tradition in the County Agricultural Boards concerning the importance of the client relationship should take precedence over national management.

The interviews I carried out four or five years after the computer system was introduced revealed that no uniform practice had developed among the forest rangers for making valuations, and in particular, for adjusting to local conditions.

The following questions were important to the forest rangers: how many of the factors on which the valuation was based should be written into the computer programme and how many should remain the province of the local case officer? Is there any scope for making local modifications to the computer system? The forest rangers were strongly in favour of this.[48] A majority felt that Per Svensson, the systems expert, was the person who wanted to meet this demand. As a result, his colleagues accused him of being disloyal. They felt that he had not discharged his responsibilities as a civil servant to national-level management, placing a different interpretation on the concept of the professional responsibility of civil servants. Per Svensson's response was to voice his lack of faith in the judgment and sense of responsibility of highly-placed staff who had no practical know-how:

A computer system in which only one person can change the conditions that apply to many other people and which have a direct effect on their work situation requires a very great degree of moderation on the part of management/senior staff. It is probable that direct measures are required to block all efforts to introduce central control over the heads of the users because the judgment and sense of responsibility of highly-placed staff members are not to be relied upon.[49]

To summarize: over time a crack appeared in the edifice of standards and agreements which had been built up among the Central Office staff. There were differences in their perceptions of reality which originated in differences in their experience. Consensus ceased to apply when differences in language and perceptions of reality appear among the members of the staff.

Maintenance of the Forest Rangers' Professional Skills

What were the actual effects on the professional skills of the forest rangers? What exactly is the erosion of professional skills? So far we have discussed a number of fears related to the possible implications of computerization and the conflict concerning what could be written in to the computer system and what could be kept open.

The shift in professional role was substantial. About five years after the computer system was introduced the forest rangers were working more as lawyers than technicians. They spent more time on land acquisition cases than before, and also on purchasing and selling, at the expense of their forest valuation work.[50]

In my interviews with the forest rangers, they repeatedly returned to the importance of manual calculations in maintaining the level of professional forest valuation skills as the following two comments serve to illustrate:

When we were doing the calculations by hand the valuation was more 'alive'. One knew what each step meant and how much weight each factor carried. Errors could be corrected. But EDP is more anonymous. Systematic errors can remain hidden.[51]

There is the issue of data-based models acting as senior supervisors' in ongoing work. New officers never learn the theoretical and practical basis on which values are calculated in the same detail as they would from manual calculations. They gradually become 'slaves' of the system.[52]

Both these comments refer to the long-term effects on building up and maintaining professional skills. Another factor that emerged in interviews was that computerization made the factors and relationships which affect the value of a property more difficult to understand. This was what the erosion of professional knowledge involved.

The Phenomenon of Duality

In a conversation that took place in 1975 between the forest rangers and Ulf Larsson, newly appointed Head of the Forestry Division, the question arose of how to follow the calculation processes, i.e. the algorithms written into the computer programme. In his reply, Ulf Larsson uses the expression 'judgment in manual calculations' and attempts to bring the discussion to a head by dividing the calculation of property value into a routine part and a more complex part. The idea behind this division was to arrive at an understanding of how to safeguard and develop the forest rangers' professional knowledge of the more complex part of their work. The forest ranger, however, chooses not to follow Larsson's line, preferring to go into the issue of manual calculation in some detail instead. His response is on a different level. He talks about manual calculation as an immediate participatory experience in which one is actively involved at every stage. He chooses sensory and physical metaphors to describe the nature of this experience: the material was more alive, it was shaped in his hands, he saw if it was out of balance. He emphasized how his judgment emerged in the course of the manual calculation and pointed out that one of the effects of computerization had been to make it more difficult to uncover errors. But at the same time he noted that people had more faith in valuations produced by the computer.

Ulf Larsson, however, did not give up the idea of dividing the work into routine and more advanced areas. He tried to bring the discussion back to his level. He wondered what the point was of having a highly-developed ability to perform calculations and asserted that it was just as important to be able to make a critical analysis of the results of the computer's calculations. Deliberately or inadvertently, he ignored the point of the forest ranger's reply – the importance of 'manual calculations' and not just 'the ability to make judgments'. Put in other words than those used by the forest ranger: when giving their views of computer support in their work, the forest rangers repeatedly returned to the point that the ability to calculate and the ability to make judgments were two sides of the same coin.[53] This phenomenon may also be seen in the responses of the forest ranger. It soon becomes clear that what Larsson called laziness in analyzing the results of the computer calculations is actually the inability to make a judgment on what one has not become familiar with.

After this, Ulf Larsson changed the subject. He talked about in-service training led by experienced and knowledgeable instructors, commenting that it took a long time to know this work and that only practical work provided experience.

Computerization broke the link between calculation and judgment. Per Svensson wanted to maintain this link: he designed the procedures at the terminal to allow the forest rangers to sit and make judgments just as they did in their manual calculations. But his good intentions were not enough.

Why was manual calculation so important? Firstly, it was not a question of purely mechanical calculations but of calculations that were interspersed with plausibility judgements. Calculations combined with judgment to form a whole. One could not therefore draw a clear line between routine and complex operations.

Again, calculation can give one a deeper knowledge of the data collected at the inventory stage. When the forest ranger uses this material in his calculations – and not until this point – he gets a total picture, an overview.

This overview emerges naturally, effortlessly; the forest rangers sees the proportions, the factors that weigh heavily, the effect a variation will have.

The overview produced by the calculation process is also important when the forest ranger makes his inventory in the field, enabling him to take into the forest, so to speak, the impression of his in-depth understanding and the results of his reflections.

Notes

1. NORDENSTAM, Tore *Sudanese Ethics* Almquist and Wiksell, 1975.
2. *Ibid.* p. 140.
3. GÖRANZON, Bo (Ed.) *Ideologi och systemutveckling* (A contribution to the discussion on science, technology and society.) *Studentlitteratur*, 1978b, p. 139.
4. We defined the terms standardized interview, occupational group meeting and taped interview as follows:

Standardized interviews: The aim of a standardized interview is to get an overview of the organization by making direct contact with different departments and individuals. A standardized interview has a number of main themes and may follow more or less freely-written interview programmes which may vary from one group of employees to the next. If it is possible for the interviewer to have direct contact with a single organization, then new problems and ideas may emerge in the course of an interview. If the organization is geographically widespread, the interview becomes more standardized and telephone contacts may be an essential complement to a circulated interview programme. When the standardized interviews have been completed, all the responses are compiled and circulated to the interviewees. The processing is primarily oriented towards qualitative characteristics, and when the responses are compiled their authentic core should be retained and any editing should be avoided as much as possible.

Occupational group meetings: Systems development in the environment of computer technology generates contacts between groups of employees (provided the systems development so permits) and promotes an exchange of information among the employees which they themselves indicate is interesting in terms of their work situation. A group may meet to discuss the standardized interviews with employees who, before the meeting, were given the opportunity to prepare themselves to discuss the responses given in the standardized interviews. Individual employees will then not be isolated with their particular problems but will be given the opportunity to discuss their problems in a given context. On the basis of the experience gained from the standardized interviews, the researcher may suggest a theme for the discussion and give examples of aspects of the chosen theme as a starting point for an occupational group meeting.

Recorded discussions: The work of structuring knowledge in systems development may present a group of employees with material such as taped discussions where key people in systems development put forward their views and tackle issues which will compel people to reflect on their own values and opinions and on the resources for development work. Wherever possible, any editing of taped discussions should retain the authentic tone of the discussion and only edit out blind alleys and irrelevant digressions in the conversation. The people involved in the conversation may discuss the editing process or, with their approval, the editing may be carried out by the researcher himself and then be made available to everyone involved in the discussion.

There is a description in GÖRANZON, Bo *Två vetenskapliga traditioner* PAAS, Report No. 6, IndEkoOrg. Stockholm Institute of Technology, 1977b, of the outline of a scientific, theoretical background to the concepts of standardized interviews, occupational group meetings and recorded discussions which were part of the method chosen.

5. There is a description of the study of computerizing the work of forest rangers in GÖRANZON, Bo *Studier av arbetsorganization och datasystemutveckling. Kunskapsuppbyggnad vid systemutveckling genom standardiserade samtal och tvärkontaktmöten i en fallstudie* (The structuring of knowledge in systems development by means of standardized interviews and occupational group meetings in a case study) PAAS, Report No. 3, IndEkoOrg, KTH. 1977a.

6. DANIELSSON, Albert *Företagsekonomi – en översikt* Studentlitteratur, 1975.

7. ANDERSSON, John (dissertation) *Industriföretagets produktionsstyrning* Ind Eko Org, The Institute of Technology, Stockholm, 1974, is influenced by Danielsson's sedimentation principle. John Andersson identifies three phases in systems development in production management in an industrial company: 'the investigation' (commonly called 'systems analysis'), 'synthesis' ('systems design') and change ('implementation'). These phases are not separable in time. The process of change starts at the time people in an organization know that a new system is to be installed. This has an effect on the existing system which means that the subject to be analyzed changes in the course of the investigation process, particularly in terms of employees' attitudes and expectations.

8. The National Agricultural Agency consists of the National Board of Agriculture and an agricultural board in each county. Its work includes the purchase, sale and exchange of forest land. There is a forestry section for valuing forest land and interpreting legislation such as the law on land acquisition. The forestry section is organised as follows:

- The head of the forestry section, whose responsibilities include the development of methods for forest valuation.
- A staff of four forest officers responsible for monitoring the market when forest land is bought and sold, and methods development.
- Fifty forest rangers, who carry out valuations of forest land and purchase, exchange and sell land. The forest rangers report to the County Agricultural Boards on this aspect of

their work, but they are responsible to the forestry section of the National Board of Agriculture in matters related to computation techniques and valuation theory.
- Approximately 25 assistant cartographers and staff who are involved in calculations perform a service function at the County Agricultural Boards.

The total area of forest sold each year is roughly equivalent to the forest bought, but the number of sales completed – about 2000 a year – is double the number of purchases made. The large number of sales is because the land is often divided up and sold to different buyers.

(See SVENSSON, Per The systems development process for a computer-based model for forest valuation, published in GÖRANZON, Bo et al.: Perspectiv data system utveckling. Om data tekniska milieu, systemutvecklings processen och arbetsorganization Studentlitteratur, 1978, p. 129.

9. See SVENSSON, Per Systemutvecklingsprocessen för en datorbaserad modell för ekonomisk kalkylering av skog published in GÖRANZON, Bo 1978, p. 130.
10. A seminar at IndEkoOrg., Stockholm Institute of Technology, 1975–09–29 with the head of the structural department at the National Board of Agriculture, on the theme of The development of forest valuation methodology unpublished transcript of a tape recording, p. 5.
11. Ibid. p. 5.
12. Ibid. p. 6.
13. Ibid. p. 7.
14. Ibid. p. 8.
15. Ibid. p. 9.
16. Ibid. p. 12.
17. Ibid. p. 11.
18. Göranzon, 1977b, p.78.
19. Per-Johan Åge, the forest officer responsible for methods development, gives the following descriptions of the main operations involved in forest valuation:
    1. Preliminary work, involving an examination of the parcel of land and the title to it, acquiring maps and aerial photographs and producing a diagram for the field inventory. The time required for this work varies between two and six hours, which is divided between the forest ranger and the office staff.
    2. The field inventory, carried out by the forest ranger. The average valuation covers about 20 sections, each section being described under two headings: class (age, site quality, degree of maturity), and measured or assessed value (volume/hectare, average diameter, felling percentage).
20. Göranzon, 1977b, p. 79.
21. Ibid. pp. 83ff.
22. Taped interviews with the head of the structure department: 1975–03–07 and 1975–03–15, and unpublished notes.
23. Ibid.
24. Ibid.
25. Göranzon 1977a, p. 64.
26. Göranzon 1977b, p. 64.
27. Ibid.
28. Ibid.
29. FORSBERG, Gun-Marie Operatörsrollen för en datorbaserad modell för skogsvärdering published in GÖRANZON, Bo, 1978a, p. 147.
30. Ibid. p. 148.
31. GÖRANZON, Bo Arbetsorganization och datasystemutveckling. Kunskapsuppbyggnad genom facklig förankring av en fallstudie, PAAS, Rapport nr. 5, IndEkOrg KTH, 1977, p. 28.
32. The office staff's situation, where they had slipped into a so-called intermittent position was discussed in the Swedish Federation of Civil Servants in September 1976. In the subsequent discussion some of the committee members commented on similarities with the following types of application:
- Information retrieval systems for physicians and lawyers.
- A planned computer system for tax computations which affects the day-to-day work of accountants.
- Information processing systems for employment services.

(See GÖRANZON, Bo, 1977c, pp. 48–49.)

Professions which act as a filter in information transfer – so-called intermittent people – secretaries, consultants, etc. – are on the increase in society. The growth of intermittent functions as a new professional group is, according to the English telecommunications expert, Colin Cherry, a probable development when new technology is introduced in working life. The relevance of Colin Cherry's interpretation is verified in the computerization of forest valuation (see GÖRANZON, Bo, 1973, p. 1).

33. Göranzon 1977c, p. 28.
34. *Ibid.* p. 30.
35. *Ibid.* p. 30.
36. Göranzon 1977a, p. 82.
37. Taped interview with Lennart Kallstenius 1975–03–07, 1975–03–1 and unpublished notes.
38. Interview with Gunnar Rosquist, a forestry officer on the staff of The National Board of Agriculture. Mr Rosquist was a colleague of Per-Johan Åge. They had the job of reviewing the method of forest valuation.
39. *Ibid.*
40. SVENSSON, Per *Tjänstemannaansvar* (The Responsibilities of a Civil Servant) unpublished manuscript, 1976–10–10, p. 5.
41. SVENSSON, *Per Om expertsystem – att veta eller att mäta* 1987–12–14, unpublished manuscript.
42. *Ibid.* p. 3.
43. *Ibid.* p. 3.
44. *Ibid.* p. 4.
45. Göranzon 1977b, p. 80.
46. See the standardized interviews with the forest rangers in Göranzon 1977a, p. 38.
47. *Ibid.* p. 68.
48. Göranzon 1977a, p. 39.
49. SVENSSON, Per: Tjänstemannaansvar (The Responsibilities of a Civil Servant) unpublished manuscript, 1976–10–10, p. 6.
50. The average age of the forest rangers was 44, and the average length of service was 15 years. The following division by type of work was noted:
    Forest valuations: 40%
    Purchase and sales: 20%
    Land purchase: 30%
    Other: 10%
    Computerization released an extra 12 hours a week to be spent on land acquisition and purchasing and selling (Göranzon, 1977).
51. *Ibid.* p. 38.
52. *Ibid.* p. 32.
53. This term is taken from Sjöberg, Alf *Brecht's 'Galilei'*, No. 42 Dramaten, 1974/75.

Chapter 3

The Dream of the Exact Language

It takes a lifetime to learn about nursing. We must make progress year by year. It takes five years, not of words but of practice, to become a nurse.[1]

Thus spoke Florence Nightingale at the age of 77 to an audience of newly-qualified nurses. The main message was her anxiety about the future of nursing.[2]

This echoes the point made by Ulf Larsson at the end of Chapter 2 concerning the future of the forest rangers:

Those who have been in this business longer will have to spend some time training new employees. . . . It takes years of practice to become familiar with this work. People gain intimate knowledge of the job by doing the practical work.[3]

In his reflections on mastering the skills of his trade, Thomas Tempte, a cabinet maker, introduced the term 'the practical intellect', partly to draw attention to the fact that complex thinking is required in so-called practical trades, and to warn against excessive faith in the abstract intellect. Tempte rebuilt a chair belonging to Tutankhamun dating back to 1350 BC, and in his reflections on the experience, asked whether the chair expresses anything that can be attributed to the person whose hands and mind made it. His answer is as follows:

What, then, is the carpenter's contribution? It is the very existence of the chair. His contribution is the expression of all the proportions of the chair and the solutions to technical design problems, but first and foremost, the time he has spent on it. Care is reflected in every part of this chair, with its well-rubbed shapes which emerged from day after day of careful work. Does the chair express anything of the person whose hands and mind made it? I maintain that it does. A very special attribute, namely a creative and imaginative intellect, was required to give material form to something which was surely subject to numerous instructions from the many people involved. I believe that the entire chair was made by using the method of fitting, that is to say the parts are put together, measured, taken apart, adjusted, and assembled once again. This method, with its mixture of repeated work, constant attention, precision and care with parts that are constantly being worn down and the risk of damage to the parts, requires a highly-developed mind. In his mind's eye the carpenter must have a clear picture of the final result and at the same time feel that each little insignificant operation has a value of its own in the context of the total work. This is mentally fatiguing. The same kind of emotional drain is experienced by a boat builder laying a clinker hull.[4]

Tutankhamun's chair (ca. 1350 BC) Photo: The Cairo Museum

Thomas Tempte was apprenticed to a boat builder for a time. His account of how the boat builder passes on craft knowledge describes how apprentices are trained to use their judgment and their minds. It also illustrates the complexity of the practical intellect.

Gösta's knowledge gained a further dimension when he became the master of so many interested apprentices. Everyone was aware of the need to pass on Gösta's knowledge and when his burden as a breadwinner was lighter, he was able to spend time demonstrating and giving explanations in conversations with his apprentices. . . . Gösta's advice to us as apprentices was to not to use tables and systems, but judgment, one's senses, training the eye to make measurements, to work on the basis of what you see and to constantly think ahead. 'All the lines should be pure and harmonious, with no bulges to offend the eye.' Cutting each plank from the timber, shaping it and fixing it into the hull is an act of birth. Gösta seems to wander aimlessly around the workshop looking at irrelevant things, standing still for long periods, making a remark about the positioning of a tool, looking at the weather. His hand-rolled cigarettes are lit and stubbed out.

The reconstruction of Tutankhamun's chair by Thomas Tempte. Photo: Thomas Tempte.

He goes out and pulls the timber about or simply looks at it. He judges, assesses and considers. Finally, he makes his decision. He asks us to help lift in a load of timber for planks. He hums and measures, makes markings with his carpenter's pencil. Then he stops for a proper smoke. It is an act of concentration. Sudden action interspersed with periods of total relaxation. But never stress. The stress and the work are kept inside Gösta. Calmly, we set out the work pieces and switch on the combination machine. He goes over to the hull, measures, assesses and checks that his thinking is correct. The width and curve are right. He makes the port and starboard pieces at the same time. Just like a cabinet maker, he makes mirror-image pieces. The measurements are finally drawn on to the batten, which is bent to give the desired curve. Fix the batten lightly in place and draw the shape. Once marked, the shape is sawn out. Another break – perhaps it is lunch time. Now the plank must be bent and put in place. That can hardly be done at once with an oak plank. The wood has to be steamed first: the plank is put in a long caulked wooden box through which hot steam is passed. This makes the plank soft and easy to mould. It is a tricky process. A piece can easily jump out of the form and it will then be unusable. It may also warp. It is difficult to describe all this in words, but in time one can learn from experience to judge the particular properties of each different plank. This is what Gösta is doing now.[5]

Thomas Tempte's description of this particularly practical trade, which requires all of our senses, is suggestive in that this dramatic and literary portrayal of work gives the account general validity. But there are also some striking, almost astonishing, parallels to be drawn between the work of boat building described by Tempte and the work carried out by forest rangers before computerization of their calculation routines. They too described the way they used their senses. In other words, the difference between the practical and the abstract intellects can to some degree be given broader content: these terms may define two different relationships to a culture within one and the same profession – two separate working cultures.

Ars Magna: Thinking as Calculation

In her study entitled *The Art of Memory*, the English intellectual historian, Frances Yates, illustrated how apprenticeship and the art of memorizing were highly-developed in ancient times and remained so until fairly recently. She demonstrates the inner relationship between apprenticeship and the art of memorizing which constitutes the practical intellect. This was a tradition in which a teacher introduced a basic idea enabling pupils to remember what they learned. The teacher gave some examples and then asked pupils to make up their own. Yates maintains that the teacher does not take 'a thousand set instructions and give them to the student to learn by heart.'[6] No technique is mastered without learning proper judgment acquired through personal experience. Only an experienced person can apply the rules which make up 'the method'.

Frances Yates shows how the emergence of the natural sciences in the seventeenth century gave 'the art of memory' a new direction. Interest in knowledge was directed at finding immutable laws, for example the laws of physics. There was a search for exact rules; mastering the rules was 'the art of memory'. This art was an important factor in the development of a scientific method.[7] On the other hand there was no interest in the practical problems of life, such as the knowledge used in different kinds of work. But there was a far greater ambition to cover all knowledge.

In his book *The Dream of the Exact Language*, the historian, Tore Frängsmyr, pointed to the Spanish monk, Ramon Lull, as a forerunner of a change in attitude towards knowledge.[8] Lull's fundamental idea, published as early as 1274, may be summarized in the term *ars magna*. *Ars magna* stated that all knowledge was built upon a number of basic elements which, when combined in different ways, could produce all possible knowledge. To show this, Lull used figures in the form of diagrams and circles which could be positioned to find the different combinations possible. He designed a rotating instrument consisting of concentric circles which immediately give the desired combinations. These instruments – described by Frängsmyr as 'both a beautiful and exciting knowledge machine' – were made of vellum or metal and the various sections and circles were painted in different colours to make them easy to read.[9]

Each subject or area of knowledge had its own figure, with a different number of circles. Frängsmyr gives an example of its use:

One of the more curious subjects was the art of preaching; by using the instrument Lull constructed for the purpose, a priest could produce a hundred examples of new subjects on which to preach. The instrument could be further refined by adding a circle with questions, divided into nine sectors under the headings which, what, from where, why, how big, when, where and how. By using these circles, Lull could illustrate every possible aspect of a problem. Sometimes he answered the questions and made comments upon them, and sometimes he did nothing more than pose the questions.[10]

Tore Frängsmyr makes the following reflection on the extent to which Lull may be said to have set the trend for the 1600s, leading to our own times.

Lull was frequently referred to as the father of symbolic logic and his thinking machine as a first, if primitive, logic machine. That may be going a little too far, but an examination of the principle reveals similarities between Lull's art and modern attempts to find the exact language. The very belief that this was possible is similarity enough.[11]

Frances Yates points to different link with our times when she underscores an idea of rationalization in the attitude to knowledge:

There is an approach to the art of Ramon Lull which regards it as a machine, which would constitute the same sort of labour-saving device in a scholastic disputation or medieval university as an adding machine in a modern bank or business office. By properly arranging categories and concepts, subjects and predicates in the first place, one could get the correct answer to such propositions as might be put.[12]

Clavis Magna

The Sixteenth century Italian philosopher, Giordano Bruno, adopted Lull's ideas and further developed his circular figures.[13] He combined this art with the tradition in classical rhetoric related to the art of memorizing. Lull wanted to create a clavis magna, a kind of universal language containing not only language but also, to all intents and purposes, thought itself.

In the seventeenth century, the philosopher René Descartes created applied mathematics as it is known today. In 1637 he presented a study in which he showed how, by applying abstract algebraic concepts, it is possible to formulate geometry's concrete points, lines, surfaces and volumes. He demonstrated a link between our three-dimensional world and a math-ematical-logical way of thinking.[14]

Throughout the seventeenth century there was vigorous debate on language as a vehicle for knowledge, a debate which was more or less shaped by the methods of mathematics and logic. The philosopher Gottfried Friedrich Leibniz was inspired by this idea. In *De Arte Combinatoria* (1666), a work he produced in his youth, he attempted to develop Lull's art along mathematical lines.[15] Despite some criticism of Lull, he adopted his method of starting with simple concepts and combining them to form more complex ones, but Leibniz's approach was less mechanical than that of his predecessor. His system was based on logical division, where single letters represented the simple concepts and combinations of letters for combined concepts. This project was further developed under the influence of his mathematical study to form a *characteristica universalis*, which used mathematical symbols and advanced methods of logic. He compared the symbols to Egyptian hieroglyphics and the calligraphic symbols of Chinese writing.[16]

Leibniz regarded language as a calculation in which the definition of concepts can have clear, sharp limits. Within the language, rules of calculation may be developed along the lines of those used in mathematics.[17] Leibniz felt that formal calculation would strengthen the intellect to the same degree as Galileo's telescope made it possible to see more clearly.[18]

Leibniz declared that

characteristica universalis . . . will constitute a new language which can be written and spoken. This language will be very difficult to construct, but very easy to learn. It will be quickly accepted by everybody on account of its great utility and surprising facility, and it will serve wonderfully in communication among various peoples.[19]

The philosopher and historian of science, Stephen Toulmin, said that the attempt to construct an exact language to 'serve wonderfully in communication among various peoples' must be judged against the background of the Thirty Years War in central Europe between 1618 and 1648.[20] 'It was a noble dream, but a dream nonetheless'.[21]

At more or less the same time as Descartes' ingeniously developed mathematical language, the mathematician and philosopher, Blaise Pascal, presented his adding machine in 1642.

I present to the general public a little machine which I have invented myself and which you can use as a tool, to effortlessly carry out all arithmetical operations and free yourself from the soul-destroying work when you use your calculator and your pen.[22]

The Finnish philosopher, Georg Henrik von Wright, says it was no coincidence that development of the calculating machine began at this time. He coined the phrase 'the theory of calculation' to describe the efforts of Leibniz and others:

What I have called the theory of calculation here is substantially the idea of the calculating machine. It is no historical coincidence that the first calculating machines were built by two of the great pioneers of New Age science and thinking, Pascal and Leibniz.[23]

It is not clear what Pascal meant by a 'calculator'. It may have been an abacus.[24] Pascal's machine, called 'La Pascaline', could add and subtract. He originally intended it to multiply and divide without any additional help. In a biography of her brother, Gilberte Pascal Perier writes as follows. 'This work (La Pascaline) was considered as a new phenomenon, having reduced to mechanism a science that exists entirely in the mind, and having found the means of doing all operations entirely reliably, and with no need for thought.'[25]

Gilberte may have been an adoring sister, but she was also an important witness of what the inventor's own opinion of his machine. The philosopher, Thomas Hobbes (1588–1679), was, if possible, even more impressed and said:

Brass and iron have been invested with the functions of brain and instructed to perform some of the most difficult operations of mind. . . . In what manner so ever there is place for addition and subtraction, there is also place for reason ... for reason is nothing but a reckoning of the consequences ... when a man reasons, he does nothing else but conceive a sum total from additions of the partials.[26]

The Effectiveness of Language

With the advent of the natural sciences came an interest in precision and efficiency. In the seventeenth and eighteenth centuries the mathematical language, with its precise, unambiguous, easy-to-use and universal symbols, was a model of standardization. The tendency towards standardization culminated with one of the milestones in the development of the English language: Samuel Johnson's *Dictionary* (1755). From a fairly comprehensive list of descriptions of words, Johnson transformed the dictionary into an authoritative national standard for the use of language. By making detailed distinctions, often with the help of quotations, he established the exact meaning of words and their correct use. His aim was to fix these meanings and usages for all time, exactly as the word triangle has meant the same thing for thousands of years.[27] But as he himself admitted, Johnson failed. Language is not responsive to such precise and definitive description; it is dynamic.

The standardization of language was followed by a critical analysis of the efficiency of language. The famous political thinker and moral philosopher, Jeremy Bentham (1748–1832), also devoted himself to this problem. Nouns are better than verbs, he said. An idea contained in a noun 'stands on solid rock' while an idea contained in a verb 'slips through your fingers like an eel'. The ideal language would be like algebra; ideas would be represented by symbols in the same way as speech is represented by letters, thus eliminating all ambiguous or imprecise words and misleading metaphors. Ideas would be linked to each other with the smallest possible number of syntactical connections, just as all numbers are linked to each other through a small number of operations – addition, multiplication, etc. Two statements could then be compared in the same way as two equations, for example, when one equation is extracted from another by multiplying by a constant. The efforts to introduce symbols for nouns and consonants, of which Bentham's work was part, were in the same spirit as Leibniz' plan to create a symbolic language. But while Leibniz attempted to make logical thinking easier, Bentham's objective was to achieve precision.[28]

The systems designer in the building company mentioned in Chapter 1 is a ghost of Bentham's thinking. When he ran into difficulties in computerizing heating design calculations, the problem for him was that the engineers did not use heating design terms with the precise definitions he had given these terms in designing the computer system. His solution to the problem was to send the heating, water and sanitation design engineers to a three-day course in systems technology: 'not to turn them into systems designers but to give them a better idea of what is important and what is less important for the systems designer.'

What is Meant by 'Interpreting' a Calculation?

Leibniz's thinking on symbolic logic was developed by the English mathematician, George Boole. His *Investigation of the Laws of Thought*, published in 1854, is frequently quoted and has become something of a classic, particularly in computer science, where Boolean algebra defines the general conditions under which combinations of statements are either true or false.[29] Here we find the background to the fact that computer design is based on two conditions symbolized in the computer as zero and one; what is called the binary system. Boole thought that the theories of symbolic logic were intimately linked to those of language. He compares this with our attempts to give exact definitions in everyday language:

It is in fact a method similar to the one we use when in daily speech we count descriptive phrases to give more exact definitions.[30]

Boole's intention was to express the laws of logical thinking in the symbolic language of mathematics, and thus establish a 'science of logic'. It was a matter of 'demonstrating how language and arithmetic serve as instrumental aids in argument', he said in the foreword to his book, but in his conclusion he felt able to hazard a guess concerning the nature of man's ability to think.[31] Boole was fascinated by the great similarities he had found between the laws of algebra and those of thinking. Algebraic notations could be applied largely unchanged.

The physicist, Pehr Sällström, who made a study of mathematical notations, says that

Boole's meticulous description was an important contribution, but Gottlob Frege was the first to really get to grips with concepts such as 'meaning', 'reference' and 'truth' in language statements.[32]

Frege, who is now often referred to as the greatest logician since Aristotle in the history of philosophy, lived a quiet and withdrawn life as a professor of mathematics in Jena, 'but happily he numbered some of the keenest brains of this century among his pupils', says Sällström, and mentions Bertrand Russell, Rudolf Carnap and Ludwig Wittgenstein.[33] Frege says he agrees with Leibniz on the value of a suitable system of notation but finds his ideas far too grandiose. He uses the following metaphor to explain his *begriffsschrift* (notation of concepts):

I think I could explain most clearly the relationship between my notation of concepts and the language of practical life by comparing it with the relationship between the microscope and the eye. By virtue of its versatility and the flexibility with which it can adjust to the most varied conditions, the eye is far superior to the microscope. Seen as an optical instrument it has demonstrable imperfections, which only remain unnoticed because of the eye's inner links with the life of the intellect. However, as soon as scientific purposes require sharp definition, the eye alone is manifestly inadequate. The microscope, by contrast, has been perfected for such purposes and is thereby unusable for other purposes. Thus, this notation of concepts is an aid developed for specific scientific purposes which should not be criticized simply because it does not work in other applications.[34]

The slim volume entitled *Begriffsschrift, eine der aritmetischen nachgebildte Formelsprache des reinen Denkens* was published in Halle in 1879.[35] A number of works appeared over the next few decades, partly inspired by Frege and

Photo: Peter Gullers

partly by independent parallel work. They came to form the basis for the work carried out in the field of philosophy in the twentieth century. With A.N. Whitehead, Bertrand Russell wrote *Principia Mathematica*, published in three volumes from 1910 to 1913. Ludwig Wittgenstein's famous *Tractatus Logico Philosophicus* was published in 1922. David Hilbert published *Grundlagen der Mathematik* in 1934 and Kurt Gödel's famous essay *Über formal*

Undenscheidbare Sätze der Principia Mathematica und verwandter Systeme was published in 1931.[36]

However, the question of what the 'game of symbols' has to do with reality, what its relevance to practical life may be, is shrouded in an impenetrable mist. 'There is – frankly speaking – no one who can properly explain what interpreting a calculation actually means,' observes Anders Wedberg in his *History of Philosophy* (1965).[37]

The Principle of the Economy of Thought

At the end of the last century the physicist, Ernst Mach, wrote an essay entitled *The Economy of Science*.[38] His thinking was a logical development of the view of knowledge as rules which can be combined in a precise way. Mach used the term 'economy of thought' to describe his ideal of science, while 'The purpose of scientific teaching is to free an individual from gaining experience by letting him take advantage of the experience of another individual.'[39]

The individual, having a relatively short life span, is at a disadvantage. Knowledge of any importance can only be gained through the greatest possible economy of thought. Science must therefore aim to provide the most complete picture possible of important facts, using the minimum amount of thought.

Experience is analyzed, divided into simpler and more familiar components and symbolized, primarily in the form of words, but preferably, and better, with ideographic symbols. Thus we represent in symbols only that part of our experience which we consider to be important – this too is an expression of the economy of thought![40]

Mach's views inspired the Vienna Circle and modern logical empiricism.

Alternatives to the Mathematical View of Knowledge in the Age of Enlightenment

We have traced a line from the breakthrough of the natural sciences up to the 1930s. Let us now return to the Age of Enlightenment, when thinkers entertained contradictory concepts of knowledge. On the one hand, their view of knowledge shows continuity with the seventeenth century and its orientation towards logic, mathematics, calculations and exact language. On the other, there are links with a different interest in knowledge, namely in the kind of knowledge that is applied in practical work.

'Bête machine', one of the earliest documents produced in the French Age of Enlightenment, was published by Doctor de La Mettrie in 1748.[41] It was Descartes who coined the phrase *bête machine*. It was also Descartes who distinguished between consciousness and body, since then a very evident part of western thinking. Descartes' project was to determine what separated man from animals. *'Bête machine'* stood for the automatic, mechanical and uniform behaviour of animals, but in his *'Discourse'* in which he presented his revolutionary mathematical theory, the word machine is applied for the first time in history to the human body:

And this will not seem strange to those, who knowing how many different automata or moving machines can be made by the industry of man without employing in so doing more than a very few parts in comparison with the great multitude of bones, muscles, nerves, arteries, veins, or other parts that are found in the body of each animal. From this aspect the body is regarded as a machine which, having been made by the hands of God, is incomparably better arranged, and possesses in itself movements which are much more admirable, than any of those which can be invented by man.[42]

Descartes continues with an important argument:

Here I specially stopped to show that if there had been such machines, possessing the organs and outward form of a monkey or some other animal without reason, we should not have had any means of ascertaining that they were not of the same nature as those animals. On the other hand, if there were machines which bore a resemblance to our body and imitated our actions as far as it was morally possible to do so, we should always have two very certain tests by which to recognize that, for all that, they were not real men. The first is, that they could never use speech or other signs as we do when placing our thoughts on record for the benefit of others. For we can easily understand a machine's being constituted so that it can utter words, and even emit some responses to action on it of a corporeal kind, which brings about a change in its organs; for instance if it is touched in a particular part it may ask what we wish to say to it; if in another part it may exclaim that it is being hurt, and so on. But it never happens that it arranges its speech in various ways, in order to reply appropriately to everything that may be said in its presence, as even the lowest type of man can do. And the second difference is, that although machines can perform certain things as well as or perhaps better than any of us can do, they infallibly fall short in others, by the which means we may discover that they did not act from knowledge, but only from the disposition of their organs. For while reason is the universal instrument which can serve for all contingencies, these organs have need of some special adaptation for every particular action.[43]

To return to the Age of Enlightenment and La Mettrie's book, Rolf Lindborg, the historian who translated La Mettrie into Swedish, discusses in his foreword to the book the way La Mettrie uses the concept of instinct.

A being which is in the power of 'instincts' is, according to La Mettrie's terminology, an animal: some people (infants, the deaf, the 'feeble-minded', everyone who doesn't understand a language) are thus 'animals'. A being who is educated so that 'understanding' is added to instincts, is a person.[44]

To La Mettrie, learning to understand a language, i.e., learning to use symbols, is to become a human being. Culture is what separates man from the animals.[45]

La Mettrie believed that thinking should turn from general abstractions to consider the concrete, the details. The models to be found in concrete examples nurtures us in a culture. According to La Mettrie, a mind that has received poor guidance is 'as an actor spoiled by provincial theaters', and goes on to say that separate states of the intellect are in constant interaction with the body.[46] La Mettrie struck a cord that was to characterize the contradictory views of knowledge that prevailed during the French *Encyclopedia* project.

In his view of language Denis Diderot, leader and editor-in-chief of the French Encyclopedia project, showed a paradoxical attitude towards various types of knowledge. On the one hand he was convinced that giving clear definitions of words would be a great step towards man's broader understanding of his condition. Logic and metaphysics would be very close to perfection if the dictionary of language were well done.[47] On the other hand, Diderot was convinced that the language people used was

only an approximation of their real feelings, that common terms always omitted the individual sense, and that disagreements about the meanings of words were not accidental, but inevitable.[48]

Diderot's disassociation from the formalization of knowledge was irreconcilable. It was a rejection of the reality which his senses showed him: empirical observations were replaced by abstract mental pictures. The great danger Diderot saw in the growth of a theoretical abstract science is that people no longer have access to the immediate experience of their senses. And for Diderot, the impressions of one's senses was the source of all knowledge.[49]

Diderot was also set apart from others by his appreciation of craft knowledge and by his implicit knowledge of practical life. Yrjö Hirn says in his biography:

Before his time there was no literature worth mentioning on the technology of craftsmen. Therefore Diderot was both compelled and willing to make personal visits to factories, question the workers, and on their dictation, make descriptions of their work methods. When it became apparent that the workers were unable to explain the methods they used, Diderot paid for models of machines to be made, which he worked himself, and which he took apart, thus to be able to give his readers an easy-to-understand description of the design and function of these machines.[50]

Diderot reflected on the concept of instinct, and he wrote an article for the *Encyclopedia* on this subject. He was interested in craft knowledge and creative ability, and asked:

How is it that mathematicians, when they have tested what is established by use and habit, have seen that this was exactly what most advanced mathematics would have preferred?[51]

This is, in itself, the beginning of an answer to the question, says the author and critic, Magnus Florin, and refers to the following passage in Diderot's letter to Sophie Valland:

a question of calculation on the one hand, of experience on the other. If the one is well-founded, it must agree with the other.[52]

Florin maintains that it is important to be receptive to Diderot's point: that there is an ability to judge and have craft knowledge which is quite simply there; reliable, instinctive. Knowledge and experience are in the body, and do not need to be constantly processed and evaluated.[53]

Diderot's article in the *Encyclopedia* on the term 'instinct' is polemic against Descartes' separation of body and consciousness, where instinct is linked to the animal's automatic, mechanical and uniform behaviour. Descartes' project was to find what differentiates man from the animal but there was no such decisive difference for Diderot. He went further than La Mettrie, who wanted to see culture – access to a language – as the most important distinction. Magnus Florin makes the following interpretation:

There are differences, exchanges and mutual dependency everywhere in life. Therefore his task in the essay on instinct is to show how this term, in its real meaning, is entirely related to experience.[54]

In line with Diderot's view, the French mathematician, d'Alembert, introduces the term instinct into an epistemological discussion in the introduction to the *Encyclopedia* (1751). d'Alembert talks about 'the instinct of a craftsman, which cannot be inherent, but can only be developed through experience.'[55]

According to d'Alembert we cannot divide all our knowledge into direct and reflective knowledge. Direct knowledge obtained at once without any intervention whatsoever of will; it finds, so to speak, the door to the intellect open and enters it without effort or resistance. It is the *tabula rasa* of Locke's epistemological explanation, the intellect as a blank board on which experience engraves its impressions. Reflective knowledge is that which the intellect acquires by processing direct knowledge and bringing together and combining different items.[56]

Mathematical abstractions facilitate the acquisition of knowledge by reflection, but are only useful if we are not limited by them. Caution befits humans with their limited vision and hence d'Alembert criticizes the misuse of algebra:

Efforts have been made to reduce even medicine to mathematical formulae: and the physicians who become obsessed with algebra have regarded the human body as a very simple machine which is easy to take apart, although it is in fact very complicated. It is remarkable to see how these gentlemen solve with the stroke of a pen problems in the fields of hydraulics and statistics on which the most advanced geometricians would be able to spend a whole lifetime. Those of us who are wiser or more cautious must content ourselves with regarding most of these calculations and vague assumptions as thought games to which nature is by no means compelled to submit.[57]

We should in no way let ourselves be impressed by the epistemological character of the mathematical sciences. They have the certainty they exhibit mainly because their subjects are so simple. The most abstract concepts, those which ordinary people see as being the most inaccessible, often provide the greatest clarity. The more tangible the properties of an object we examine, the more obscure our ideas become. Even the nature of movement is an enigma.[58]

In the next stage of the discussion on the development of knowledge, d'Alembert takes up what he calls the science of logical thinking. He sees this as the key to all knowledge. The art of logical reasoning is a gift nature imparts to geniuses, and it may be said that books dealing with this subject are only useful to those who can manage without them. Sound conclusions were reached long before logic, reduced to principles, taught man to analyze their modes of thinking.

The valuable art of arranging ideas in a suitable order and thereby of facilitating the transfer from one idea to another gave us, to some extent, a means of approaching what appear to be the most different of people. All our knowledge actually goes back to our perceptions, and everyone has more or less the same perceptions. The art of combining and bringing together direct ideas actually adds nothing more to these ideas than a more or less exact order and a way of arranging them that may make them more or less perceptible to everyone.

A person who has a facility for combining ideas is no different from one who combines ideas with difficulty. They are no more different than a person who immediately on seeing a painting passes judgement on it and a person who must study all the different facets of a picture one after another before he can appreciate it. When these two people look at the picture they have the same perceptions, but these perceptions have not, so to speak, been integrated by the second person. He needs to stop and fix every impression of his senses for a longer time before he can reach the point that the first person arrived at immediately. In this way the first man's reflected ideas are just as available to others as direct ideas.[59]

There are few things which cannot be reduced to simple concepts and arranged in such a close relation to one another that a single, unbroken chain is formed. As d'Alembert puts it:

The different speeds at which the mind works gives us different degrees of dependance on this chain of ideas. The advantage of the greatest geniuses is that they need to link their ideas less than others, or rather that they form a chain of ideas more quickly, and almost without noticing it.[60]

Thus, d'Alembert who distanced himself from the uncritical use of mathematics, such as the application of algebra to medicine, says something quite different in this clearly-formulated description of what he calls the

Efforts have been made to reduce even medicine to mathematical formulae: and the physicians who become obsessed with algebra have regarded the human body as a very simple machine which is easy to take apart, although it is in fact very complicated.

d'Alembert: *Introduction to the Encyclopaedia*. Volvo, Gothenburg, 1942. Photo: Peter Gullers.

Fantasy is no less active in a creative geometrician than in an imaginative poet.

d'Alembert: *Introduction to the Encyclopaedia.* Volvo, Gothenburg, 1981. Photo: K. W. Gullers.

science of logical thinking. His comment still applies to the most advanced technological development of today, so-called expert systems in the field of research into artificial intelligence.

Rhythm in Work: Thinking as Doing

Denis Diderot emphasized that the greatest mistake in philosophical thinking was the tendency to look for abstractions for all phenomena. He repeats in various contexts the need to give examples, observing that in responding to an abstract dissertation, one asks for examples:

you are really asking that the speaker give body, form, reality and some specific notion to the successive sounds he has uttered, by linking them to experienced sensations.[61]

He attempted to classify the range of expressions, gestures and mime used in discourse, not in order to put them into words, but trying rather to provide a pictorial basis for thought. Take the following quote:

We have a precise idea about something; it is present and clear in our minds, but when we try to find a way of expressing it, we find nothing. We combine words – gentle and harsh, fast and slow, soft and loud, but the spider's net is always too weak. Everything escapes.[62]

Language cannot completely reflect our thoughts. Diderot was interested in rhythm, co-ordination and movement in work. The intellect develops

through contact with sensations. Rhythm is created from the struggle with
words and expressed through variations of pause and activity. Rhythm is the
sign of a personal signature in work. Rapid movement of thought, swift
action, is characteristic of professional knowledge. The expression of rhythm
is a sign that a technique has been mastered.

What is rhythm? The musical imitation of rhythm was not an unprocessed expression of
sensations. It was rhythm created after an agonizing struggle with words. The mind listens
attentively to its own movements, the critical intellect is, always lucidly, always actively,
considering sensations which show them naturally in the spontaneous gestures of thought.[63]

Compare this with Thomas Tempte's description of Gösta the boat builder's
rhythm in his work (see above). This is where the suggestive aspect of his
description is largely to be found: rhythm is expressed in the physical action
alternating between activity and pauses in the work. 'But never stress. The
stress and the work are kept inside Gösta.'

    According to Diderot, the ability to observe and control the surge of one's
own thoughts, to make the proper gestures and rhythmic expressions, is a
result of reflection. The point of Diderot's study of the paradox of the actor is
that the actor is not subordinate to the emotion he is projecting. Actors who
play from their souls, who are sensitive, are never consistent in playing their
parts.

It is quite different with the actors who act after thinking and after studying human nature, and
who always have some ideal picture to follow, building their part on the power of their
imagination and their memories.[64]

This is an application of Frances Yates' art of memorizing and Thomas
Tempte's practical intellect.

Diderot's Paradox

In his dialogue of language and gesture entitled *Rameau's Nephew*, Diderot
establishes a distance to the two roles of dialogue.[65] These roles are 'I', the
representative of logical, calculable common sense, and 'He', the vulgar
bohemian at the bottom of society who has an affinity with the deeper layers
of sensitivity in his personality. Diderot examines what keeps these two
different characters together. 'He' attacks 'I' for retreating into abstractions
and personal isolation from a complex reality.[66] 'I' criticizes 'He' for becoming
excited over pantomime pranks, while not being able to exercise any practical
skill in his music.[67]

    The philosopher, Allan Janik, has made a thought-provoking interpretation
of *Rameau's Nephew*, where his thinking focuses largely on Diderot's ability to
portray different mental attitudes through dialogue:

Rameau's Nephew is a trenchant analysis of the relationship of teleology to instrumentality
inasmuch as it sets 'knowing that' against 'knowing how' and in so doing challenges a central
tenet in modern philosophy, namely that practical knowledge ultimately has a theoretical basis –
a view which we find reflected in doctors who refuse to admit that nursing skill is really
knowledge at all. In Diderot's view the practical and the theoretical, the experiential and the
propositional sorts of knowledge are not only radically incompatible with one another but
actually in competition with one another in the social world. However, the importance of the

dialogue hardly ends there. Diderot brilliantly juxtaposes two ways of receiving knowledge and life itself not simply in such a way that they are questioned, ridiculed or rejected, but in such a way as to show how they problematize one another.[68]

The contradictions are not resolved and it is agreed that there is a gap which cannot be bridged. The brilliance of *Rameau's Nephew* lies in the fact that Diderot does not take sides in the struggle between the senses and the intellect, but retains the complexity and the contradictory essence of the interaction between the different layers of one's own person. This may be seen as a portrayal of the paradoxical view of knowledge in the *Encyclopedia* project. The rhythm between the controlling sense and the chaotic is a balancing act which interests Diderot. Professional knowledge is not developed by means of methods and instructions, he maintains. It is developed and deepened by a great deal of practice. Reflection becomes vital: 'I am forced to reflect. It is an illness which has to run its course.'[69]

To conclude by returning to the forest rangers and forest valuation. The phrase 'economy of thought' – which the physicist, Ernst Mach, introduced to describe the purpose of science – is an accurate heading for one of the contradictions which beset the county agricultural boards: the idea that the job could be made more efficient by transferring the calculations from the forest rangers to, in the first instance, the office staff, and then to computers.

Another link may be made with the view Diderot expressed on the internal relationship between well-founded experience and calculation, the mathematical model.

A question of calculation on the one hand and of experience on the other. If the one is well-founded, then it must agree with the other.[70]

The third link is with the apprenticeship form of training mentioned at the beginning of this chapter. Jean-Jacques Rousseau, one of the most prominent figures in the *Encyclopedia* project, was interested in the relationship between different kinds of knowledge. He said that the concepts of theoretical knowledge must be traced back to a reality that our senses can perceive:

Of all our senses, sight is the sense which we can least separate from the judgments of our intelligence; and that is why it takes so long to learn to see. It takes time to compare sight and feeling, and train the former to give a true picture of shape and distance. Without feeling, without continuous movement, the sharpest eye in the world cannot give us an idea of space. To an oyster, the world must seem to be a point, and it would not seem otherwise even if it had human intelligence. It is only by walking, feeling, estimating and measuring the dimensions of objects that we learn to perceive them properly; but, on the other hand, if we were always measuring, our senses would rely upon that means and never achieve certainty.[71]

Rousseau emphasizes the importance of the apprentice Emile receiving his training through concrete work in a meaningful occupation and in close proximity with other people. Thus, according to Rousseau, training requires the teacher to give his pupil the basis for practical and reflective knowledge. This is part of a learning process which aims to bring about a process of formation whose social, ethical and aesthetic content are an essential preparation for life at work and in the community. Rousseau underlines the importance of work providing an opportunity for people to reflect on both the value of what they generate in terms of concrete production and also on the social, moral and ethical values that are an inherent part of their work as a way of life. All work leads to something greater than itself. This too is an important aspect of the value and purpose of work.[72]

Notes

1. The short quotation from Florence Nightingale's lecture is taken from her book, *Notes on Nursing, What is and what it is not*, published in 1859, (new impression, Duckworth, 1970).
2. JOSEFSON, Ingela *Från lärling till mästare*, FOU report 25, SHSTF/Studentlitteratur, 1988, p. 25. Florence Nightingale is a model in Ingela Josefson's study of the view of British nurses of the use of advanced computer technology in nursing.
   See also Ingela Josefson: Language and Experience, published in Bo Göranzon and Magnus Florin (eds.): *Artificial Intelligence, Language and Knowledge. On Education and Work*, Springer-Verlag, London, 1990a..
3. Once a year the forest rangers have a week-long meeting to deal with current forestry problems. This meeting also allows forest rangers from different counties to compare their professional experience. In 1975 a national level continuation course was held which I had the opportunity of attending. Among other things, the ideas behind a new forestry law were presented. This legislation required new models and valuation principles to be developed. Plans for rationing had to be introduced into the current forest evaluation method. The basic issue was how this would affect the forest rangers' feeling for their jobs as valuers: 'There is no reason to argue about details before you have a feeling for the total picture' was one forest ranger's comment. (Quote from ULRICEHAMN *National Continuation Course on Current Forestry Problems* 1975-10-27-30, unpublished manuscript, p. 1).
4. TEMPTE, Thomas: The Chair of Tut anck Amon, published in GÖRANZON, Bo and FLORIN, Magnus *Dialogue and Technology. Art and Knowledge*. Springer-Verlag, London, 1990b.
5. TEMPTE, Thomas *Arbetets Ära. Om hantverk, arbete. Några rekonstruerade verktyg och maskiner.* Arbetslivscentrum, 1982, p. 85.
6. YATES, Frances *The Art of Memory*. Ark paperbacks, London, 1984, p. 4.
7. *Ibid*. p. 368.
8. FRÄNGSMYR, Tore 'Drömmen om det exakta språket, published in Gunnar Eriksson, Tore Frängsmyr and Magnus von Platen: *Vetenskapens träd*, Stockholm 1974, p. 319.
9. *Ibid*. p. 324.
10. *Ibid*. p. 324.
11. *Ibid*. p. 327.
12. YATES, 1984, p. 364.
13. FRÄNGSMYR, 1974, p. 326.
14. KLINE, 1968, p. 152.
15. TOULMIN, Stephen *The Dream of an Exact Language* in Göranzon and Florin, 1990b.
16. FRÄNGSMYR, 1974, p. 328.
17. VON WRIGHT, Georg Henrik *Logik, filosofi och språk* Aldus, 1971, p. 40.
18. SÄLLSTRÖM, Pehr *Matematisk notation* SALFO/FRN, 1989, p. 15.
19. TOULMIN, 1990.
20. *Ibid*.
21. *Ibid*.
22. KLINE, p. 286.
23. VON WRIGHT, 1971, p. 43 *Ibid*.
24. The abacus was known in ancient Greek and Roman times. Stone spheres were moved from a flat stone or metal surface. The words calculation, calculator etc. come from the Latin word, 'calculus', meaning a limestone ball. Instead of a plate the Greeks used a sand box in which they could both calculate by marking the position of the balls and write down the results. The Greek for sand box is 'abax' from which comes the word abacus. The Arab culture spread the use of the abacus to Russia and further afield. This very important advance in mathematics came to Europe in the ninth century. The abacus reached China in the twelfth century, where it was called the suanpan. A century or so later it was introduced in Japan where it was called the soroban. The abacus may be seen as a first step towards the automation of mathematical operations. It is still widely used in China and Japan.
   COOLEY, Mike *Computer Aided Design – its nature and implications* TAAS Publication, 1973, p. 9.
25. STEIN, Dorothy Think Tanks, *The Guardian* 26th March 1987.
26. *Ibid*.
27. KLINE, p. 223.
28. *Ibid*.

29. SÄLLSTRÖM, p. 156.
30. BOOLE, George *Matematisk analys av logik* Sigma Vol.5, Forum, 1960, p. 1953.
31. SÄLLSTRÖM, p. 156.
32. *Ibid.*
33. *Ibid.*
34. *Ibid.* p. 158.
35. *Ibid.* p. 157.
36. *Ibid.* p. 164.
37. *Ibid.* p. 168.
38. MACH, Ernest *Vetenskapens ekonomi* Sigma, Band 5, forum 1960, p. 1883.
39. *Ibid.* p. 1883.
40. *Ibid.*
41. LINDBORG, Rolf: *Maskinen, människan och doktor La Mettrie*, Doxa, 1983.
42. *Ibid.* p. 25. See also Reneé Descartes: *Avhandling om metoden*, Idéhistorisk läsebok, Band 1, Gidlunds, 1982, p. 282.
43. LINDBORG, p. 25.
44. LINDBORG, p. 82.
45. *Ibid.* p. 80.
46. *Ibid.* p. 113.
47. MASON, John Hope *The Irresistible Diderot*, Quartet Books, London, 1982, p. 11.
48. *Ibid.* p. 11.
49. SÄLLSTRÖM, Pehr Diderot, Goethe och naturvetenskapen, published in *Dialoger* No. 13/89, Encyklopedi, p. 10.
50. *Ibid.* p. 11.
51. Diderot's article on *Instinct* was published in *Dialoger* No. 13/89 translated into Swedish by Jan Stolpe) pp. 15–17.
    The quote is from FLORIN, Magnus *Michelangelos kupol*, published in Dialoger 13/89 Encyklopedi, p. 14.
52. *Ibid.* p. 14.
53. *Ibid.* p. 14.
54. *Ibid.* p. 14.
55. d'ALEMBERT: *Inledning till Encyklopedin*, translated by Jan Stolpe, Carmina klassiker, Uppsala, 1981, pp. 31–32.
56. *Ibid.* p. 28.
57. *Ibid.* p. 50.
58. *Ibid.* p. 53.
59. *Ibid.* p. 60.
60. *Ibid.* p. 60.
61. JOSEPH, Herbert *Diderot's Dialogue of Language and Gestures*, Ohio University Press, pp. 8ff.
62. *Ibid.* p. 12.
63. *Ibid.* p. 15.
64. DIDEROT, Denis: *Skådespelaren och hans roll*, Prisma, 1963, p. 19.
65. DIDEROT, Denis: *Rameus brorson*, translated into Swedish by Ria Wagner, Tiden, Stockholm, 1951. See Horace Engdahl Dialog och upplysning in *Dialoger* 3/86, Dialog och upplysning.
66. *Ibid.* p. 41. See also JANIK, Allan Offenbach – konsten mellan monolog och dialog published in *Dialoger* 4/87, Dialog och upplysning.
67. *Ibid.* p. 42. See STOCHOYEV, Vladimir Mozart och Salieri, published in *Dialoger* 5/87, Artificiell Intelligens.
68. JANIK, Allan Caliban's Revenge, published in GÖRANZON, Bo and FLORIN, Magnus (eds.): *The Diderot Project. Perspectives and Implications*, Working report No. 1, Swedish Centre for Working Life, 1990c.
69. DIDEROT, 1951, p. 89.
70. DIDEROT *Brev till Sophie Valland*, translated into Swedish by Olof Nordborg, Atlantis, 1987, letter of 2nd September 1762, pp. 192–196.
71. ROUSSEAU, Jean-Jacques *Emile*, J. M. Dent & Sons, Ltd., 1982, p. 107.
72. DAHL, Eva-Lena Synen på kunskap i Jean-Jacques Rousseau's ideologi, in *Dialoger* 1/86, pp. 40–53.

Chapter 4

Literature, Language and Learning: Turing's Paradox and the Metaphor of Caliban

Research Method: Art as a Source of Knowledge

One should not look for an historical portrait of Galileo, but for an analogy based on the contradictions which still exist within us, and to strive to overcome them. . . . The play begins with the potential of the instrument that was to hand – it begins with the telescope. And the telescope is the instrument which makes it possible for Galileo to think and see as he does; this shows that we cannot exclude the instrument from our calculations.[1]

The director, Alf Sjöberg, attempted to interpret the epistemological effects of the picture of the world proposed by the modern natural sciences. He wished to produce on the stage a picture of the existential issues raised. Despite all the precise technical instruments which help us to make ever more exact definitions, there is always something which remains indefinable. This is the paradox of critical knowledge. Aware of this, Sjöberg wanted to see drama as a means of 'burning your way through reality to reach an inner truth', even if the answer thus arrived at is only a tentative one.[2]

Sjöberg produced Brecht's *Galileo* in 1975. He described his interpretation of the play in an introductory lecture at a symposium on the theme of professional ethics held in March 1975.[3]

The symposium was part of the on-going case study of the County Agricultural Boards; there was still serious disagreement on the issue of the responsibility to be borne – in this case – by the forest rangers as civil servants. The purpose of the symposium was to create a forum for reflection on this matter. Sjöberg's example opened the door for reflections on people's moral reservations about the development of computer technology; in particular about the question of professional ethics which comes up when the work of various occupational groups is computerized.

The discussion at the symposium centered round the issue of what a systems expert should do when he discovers that the proposals for rationalization which he has been commissioned to produce will make many people's jobs worse than they are today, and that the work force had been unanimously against the system ever since they found out what its effects would be.[4] A variety of backgrounds and disciplines were represented in the discussion at the symposium: there were people from the theatre, philosophers, lawyers, economists and representatives from working life. The discussion did not result in any general agreement, but Sjöberg's example of Brecht's *Galileo* stimulated a dialogue in which views on the points of disagreement were formulated and ideas for further reflection were discussed.[5] As a result, more work was carried out on defining the relationship between art and science. We saw this as a step in the process of reflecting on the more profound and long-term aspects of computer processing, for example, changes over time in professional knowledge and professional ethics. A symposium was held two years later, at which the following questions were raised:

Are there things which the arts – some arts – can express that science cannot? Can art be given a place in pluridisciplinary work, helping to fill the space in fields that calculating science cannot reach?[6] This symposium was on the theme of art and social change. In summing-up, attention was drawn to two very different views of the relationship between man and machine. One is an external perspective which describes technological change within the conceptual world of the natural sciences. The other is an internal perspective applying to people connected with technology. Art has a function here – to lift out and render visible the phenomena which require a more plastic language to become accessible for reflection.[7] In other words, the language of art can be of help in the study of the long-term aspects of the relationship between man and machine.

The present chapter examines this double perspective in greater detail, beginning with the way in which the myth of an exact language reached a climax in the dream of reproducing all forms of human knowledge with the help of a machine, i.e. in artificial intelligence. The chapter continues with a discussion of the heuristic and pedagogical function of art, its value as a source of knowledge with which to understand human actions.

The Turing Machine

In 1936, Alan M. Turing, the English mathematician, wrote an article which has since become a classic. This article, entitled 'On Computable Numbers, with an Application to the *Entscheidungsproblem*', contains the basic mathematical theory for computing machines.[8]

The term 'problem of decidability' – '*das Entscheidungsproblem*' was introduced by David Hilbert. It refers to the question whether there is a clearly defined method which will determine whether mathematical assertions are provable. The method must be a predictable mechanical process which may be applied independent of human judgment or choice.[9]

Turing's work in 1936 was a contribution to the field of the mathematical theory of proof, but his theorems and symbols made it accessible only to experts in that specialist field. The following is a quote from a play by Hugh Whitemore called *Breaking the Code*.[10] It is based on documentary material of Turing's life.[11] In this speech from the play, the figure of Turing talks about a kind of machine which reads off mathematical symbols and processes them mathematically, in short, the principle of the computer:

Galileo working on a new systems theory. Toivo Pawlo in Alf Sjöberg's production of Brecht's *Galileo*. Photo: Beata Bergström.

Thinking, Theatre, Transformation

Here I sit in the company of scientists and talk about the problems of the theatre. I believe that, let us say twenty or twenty-five years ago, no sensible scientist would have dreamt of inviting a director to talk about his experience. This is evidence of the living dynamics in the problems posed by Ibsen, Strindberg, Shakespeare, Sophocles, Brecht. Why have these gentlemen, those who have pushed back the boundaries, who have made great leaps forward – the mutations, who have transformed systems and people's way of thinking, why have they devoted themselves to the theatre? It is because they know that they have a dynamic force in their hands, a force which will be developed in new series, in new transformations through the passage of time.

(Alf Sjöberg, from the discussion at the symposium on professional ethics, 24 March 1975, following his lecture on 'Galileo and the Freedom of Research', in *Ideology and Systems Development*, ed. Bo Göranzon, 1976.)

Gödel's theorem is the most beautiful thing I know. But the question of decidability was still unresolved. Hilbert had, as I said, thought that there should be a single clearly defined method for deciding whether or not mathematical assertions were provable. The decision problem he called it, the Entscheidungsproblem. In my paper on computing numbers I showed that there can be no one method that will work for all questions. Solving mathematical problems requires an infinite supply of new ideas. It was, of course, a monumental task to prove such a thing. People had been talking about the possibility of a mechanical method, a method that could be applied mechanically to solving problems without any human intervention or ingenuity. Machine! – that was the crucial word. I conceived the idea of a machine, a Turing Machine, that would be able to scan mathematical symbols – to read them if you like – to read a mathematical assertion and to arrive at the verdict of whether or not that assertion is provable. My idea worked. It was a machine of the imagination, like one of Einstein's thought experiments. Building it wasn't important; it's a perfectly clear idea, after all, (and it has given me the opportunity to reword Hilbert's question like this: is there a machine of this kind – a Turing machine – which can test all mathematical assertions? And the answer is, of course, no, as is very easily demonstrated.)[12]

In the spring of 1947, Alan Turing met Norbert Weiner, the founder of cybernetics, who had the most imaginative strategy for realizing Turing's ideas.[13] Five years earlier, Norbert Weiner had published an article which marked the foundation of cybernetics.[14] The main point of this article was that the principles of comparative analysis which are applied in examining human behaviour can also be applied to machines. However, according to the authors, a functional study reveals major differences: we need not ascribe to machines characteristics such as consciousness or motive, or intentions which are identical with the human constitution. This is an important point to remember.[15]

The theory of biological development is a model in Norbert Weiner's studies. Cybernetics stresses the relationship between animal and machine and emphasizes that the particular way a machine works gives an indication of its expected performance. Wiener emphasizes the aspect of continuity in biological development. The biological individuality of an organism seems to lie in a continuity of processes and in the organism's memories of the effects of its earlier development. The process of biological development may be regarded as a chronicle of problem-solving.

Unlike machines and animals, human culture is characterized by its access to language. As mentioned in Chapter 3, Descartes, and later La Mettrie, drew attention to this fact. But Weiner points to Leibniz as the most important model for his research programme in cybernetics. 'If I were to choose a patron saint of cybernetics (the study of living organisms as logical machines) from the history of knowledge,' wrote Norbert Weiner in 1948, 'I would choose Leibniz . . . his idea that *calculus ratiocinator* (the calculating sense) contains the embryo of *machina ratiocinatrix*, the sensible machine.'[16]

In his book *La cybernétique*, the French mathematician, G. T. H. Guilbaud, developed the discussions on the significance of feedback in terms of the development of knowledge.[17] Here, Guilbaud uses Descartes' metaphor of the hiker who got lost in a maze.

There is a fairly comprehensive body of mathematical literature on the problem of extricating oneself from a maze the design of which is unknown. In this situation the relationship between knowledge and action is unusually clear; I want to reach the goal, the exit, but it is not until I walk around and explore the maze that I build up knowledge, and the knowledge I gradually acquire will dictate my future actions.[18]

This quote illustrates the inherent attraction of some of the arguments and concepts of cybernetics. In the 1950s, cybernetics was the door which opened onto the research field of 'artificial intelligence'.[19]

Turing – The Paradox

In his work on the Entscheidungsproblem, Turing placed a limit on what could be done with computers. But, paradoxically, fifteen years later he expressed unreserved faith in computers. In an article entitled 'Computing, Machinery and Intelligence', published in 1950, Turing stated that it was his conviction that computers should be able to imitate human behaviour perfectly and that this goal would be attained by the year 2000.[20] This article became a manifesto for a group of computer researchers in the field of artificial intelligence.

The English mathematician, Andrew Hodges, makes the following comment on the paradox of Turing: 'Given Turing's original interest . . . perhaps it is surprising that he himself made little of his own discovery about this absolute limitation. On the contrary, his interest focussed more and more on exploring what a machine could do, not what it could not.'[21]

One of Turing's fundamental themes is the computer as a universal machine, which Whitemore has Turing explain in the play *Breaking the Code*:

It's true computers are often used to do calculating because they calculate very quickly – but computer programs don't have anything to do with numbers. A colleague of mine has got our computer to hum tunes – it once sang 'Jingle Bells'. We've even got it to write love letters! Doing calculations, humming a tune, writing love letters. These are very different tasks, but they're all done by one machine – and that's an extremely important fact about computers. A computer is a universal machine and I have proved how it can perform any task that can be described in symbols. I would go further. It is my view that a computer can perform any task that the human brain can carry out. Any task. Now you might think from what I've said that a computer can only do what it's told to do. Well, it's true that we may start off like that – but it's only the start. (We can set the computer to modify its instructions in the course of the process.) A computer can be made to learn.[22]

Turing saw the computer as a machine which obeyed a number of rules. In his view, the most important part of the computer was the 'book of rules', i.e., the software. If one wants to make a machine imitate human behaviour in some complicated operation, one must ask the human how the operation is carried out, and translate the replies into a computer programme. 'This is usually described as programming', says Turing.[23] He maintains that it is impossible to create a set of rules which describes the way a person would behave in all conceivable circumstances. Turing introduces a distinction between the concepts of 'rules of conduct' and 'laws of behaviour'. He uses the term 'rules of conduct' to refer to rules such as 'stop when you see a red light', rules which we can act upon and be aware of. He uses the term 'laws of behaviour' to mean the natural laws of the human body, such as 'if you pinch him he will scream'. In Turing's opinion, this kind of 'law of behaviour' is an important area of research which is a theoretical basis for automating human behaviour.[24]

The meaning Turing gives to the term 'rule' has its historical roots in the idea of a universal language as formulated by Leibniz. It has a normative function, being a criterion of what can be accepted as knowledge. In his classic article, 'A General and Logical Theory for Automatons', John von Neumann agrees with Turing.[25] The article has the same epistemological approach, namely that total automation of human behaviour is possible. The quote below is reminiscent of d'Alembert and his idea that people only differ in respect of the speed with which they process the impressions of their senses.

There is no doubt that every particular phase of every conceivable form of behaviour can be described 'completely and unambiguously' in words. The description may be a long one, but it is always possible to make. To deny this would imply recognition of a form of logical mysticism which is undoubtedly foreign to most of us.[26]

Turing's Man

The 1950 article presented a method of defining 'intelligence' – the so-called *Turing Test*. A person is placed in one room and a computer in another. Both are able to communicate with the outside world but only through typewritten texts. Another person is placed in a third room and, after questioning both 'intelligences', has to decide which of them is human. Turing maintained that if the interrogator fails in his task then one must ascribe intelligence to the computer.[27]

David Bolter develops the implications of Turing's perspective, which is most trenchantly expressed in the *Turing Test*. Bolter has written a book called *Turing's Man*, sub-titled *Western Culture in the Computer Age*. To Bolter, the interesting thing about Turing's work is that it is an expression of the importance of computers in our time. The computer is giving us a new definition of man as an 'information processor', and of nature as 'information to be processed'. Bolter calls people who hold this view of man and nature 'Turing's men'.[28]

In the example of the computerization of heating, water and sanitation design engineers' calculations in Chapter 1, the systems expert appears as a 'Turing's man' in practical work. His thinking follows a simple logic: the engineers make calculations which can be computerized. This will make all the heating, water and sanitation engineers redundant, except for a few people who will be given the task of developing the calculation methods – which will then be transferred to the computer.

The human brain has always been likened to the technology of the time. The brain has been described as a telephone network, as a telegraph and, today, as a computer. What else can it be? In an article on the epistemological foundations of 'cognitivism', the American philosopher, John Searle, criticizes the choice of these metaphors.[29]

Cognitivism has close links with the idea of man as an information processor: the notion that a study of computers will teach us something about the processes by which humans solve problems.[30] The 'behavioristic trick' (cf the *Turing Test*) is to define intelligence in terms of behaviour. The ability to imitate a specific piece of behaviour is said to be 'intelligent'.[31]

It is the intellectual position of the *Turing Test* which Joseph Weizenbaum attempts to tackle in his now classic *Eliza* computer programme, which was intended to be a pedagogical example. He wanted to make people more aware of the limitations of computers by developing software which simulates a psychotherapist's questions and responses.[32] Weizenbaum found reaction to the programme totally unexpected. The psychoanalysts directly affected by the application were enthusiastic. They saw the possibility of acquiring an instrument to shorten waiting times for patients in psychiatric care:

If the Eliza method proves beneficial, then it would provide a therapeutic tool which could be made widely available to mental hospitals and psychiatric centres suffering a shortage of therapists. A computer system could deal with several hundred patients an hour.

A human therapist can be viewed as an information processor and decision maker who is applying a set of decision rules which are closely linked to short-range and long-range goals. He is guided in these decisions by rough empirical rules telling him what it is appropriate to say and not to say in certain contexts.[33]

Weizenbaum's application demonstrates that the *Turing Test* is not a satisfactory one. It can be misleading. This gave Weizenbaum an insight into a fundamental problem: human beings are liable to attribute to new technology, in this case a diagnostic programme in the field of medical care, more intelligence than it possesses. We lose our distance. We fail to realize what the limitations are. This is a feature of the emergence of a technological culture.[34]

Theatre and Technology

Weisenbaum says that he chose to name his programme *Eliza* for like Eliza in George Bernard Shaw's *Pygmalion*, the computer programme could learn to 'speak' better and better.

Joseph Weizenbaum was interested in the tradition of the value of literature in understanding the development of society. However, like Norbert Wiener, who shared this interest, he did not go on to examine the heuristic and pedagogical functions of art – its value as a source of knowledge in understanding human actions.[35]

This brings to mind the penetrating analysis of the Pygmalion metaphor by the English author and literature researcher, Julian Hilton, in his article, 'Theatre and Technology: Pygmalion and the Myth of the Intelligent Machine'. In this provocative article, Hilton introduces the idea that the most advanced area for the 'simulation of behaviour' was the theatre, which should therefore become a research area for artificial intelligence. It is possible that the particular nature of the theatre's form of intelligence can provide us with some clues as to how the human ability for thought and creativity may be reproduced satisfactorily in machines, or if not, demonstrate the impossibility of such a situation.

What AI attempts to achieve is in practice to copy in a machine the properties of the human brain in terms of common sense and imagination, which presupposes that we have knowledge of these characteristics. But the fact is that in this respect science does not know very much, while the theatre, albeit more implicitly than expressly, has for thousands of years examined the problems related to the reproduction or simulation of one person's behaviour through another's. AI may learn a great deal from examining the theatre's extensive investigations of the nature of human intelligence.[36]

And so to Hilton's interpretation of the Pygmalion myth. The play is about a language expert, Professor Higgins, who wagers that by using his expert knowledge of languages he can transform Eliza from a Cockney to a countess in six months. Higgins' assumes that if he can teach Eliza to speak as he does, she will also think as he does. The turning point in the play comes when Eliza realizes that she is fully capable of playing a countess, yet she is not a countess. At the same time she understands that her newly-learned expert linguistic knowledge has also deprived her of another kind of knowledge.

I sold flowers, I didn't sell myself. Now you have made a fine lady of me I am not capable of selling anything else. I would much rather you had left me where you found me first.[37]

Hilton asks whether this will be the fate of users of expert systems, whether using such systems will not cause a loss of certain kinds of knowledge.[38]

Literature, Reflection and the Theory of Knowledge

The Pygmalion myth is not about the use of technology – but Julian Hilton uses it to pose questions on that subject, and corroborate what the philosopher, Alan Janik, says about literature, namely, that literary works may be of vital importance in areas such as working life where, at first glance, they have no relevance.[39]

This idea is taken from an essay by Alan Janik on the theme of 'Literature, Reflection and the Theory of Knowledge', in which he asks whether there are several forms of knowledge which cannot be reduced to each other:

The rejection of this Cartesian-positivistic perception of the uniformity of knowledge is the great triumph of the philosophy of science in the last 25 years – even though there are still people who have not understood its real power and importance. I refer to the pioneering work carried by, for example, Kuhn, Winch, MacIntyre, von Wright to name only the most prominent. They paved the way for the idea that there is a form of knowledge, reflection, which cannot be effectively based on the kind of direct communication that is typical of scientific research reports. The reason for this is briefly that scientific reporting aims to bring about solutions, while reflection aims at understanding why we have the problems we have. The central realization here implies epistemological pluralism.[40]

Alan Janik chooses four literary examples – from many – to demonstrate the part which literature may play in the epistemology of reflection. Janik's fundamental idea is that without reflection on literature the basically problematic character of human experience can only be researched in a shallow, superficial way. His four examples are: *Orestia* by Aeschylos, as an example of essentially conflicting forces; Shakespeare's *King Lear* as an example of a wrongly-structured question; Diderot's *Le Neveu de Rameau* as an example of the problem of dialogue despite conflicting perspectives; and the poetry of George Trakl as an example of reflection on the ability of language to constitute the world.[41]

The Metaphor of Caliban in Our Technological Culture

In my own research, I have used literature as a source of knowledge in order to make possible reflections on phenomena which are not accessible in any other way.[42] One such approach is the initial example of Brecht's *Galileo*, which formed the point of departure for a discussion on professional ethics. Here I have chosen the metaphor of Caliban from Shakespeare's last play, *The Tempest*, because I see in it more clearly than in any literary text the paradoxical relationship between theory and practice. What happens in the meeting between well-founded experience and the formalized algorithms of theory?[43]

The meeting of Prospero and Caliban personifies a contrast between the abstract and the practical intellect.[44] They become dependent upon one another. What happens in the exchange between these two characters may be seen as an underlying metaphor for the phenomenon of professional knowledge in our technological culture. At the same time, it is a deeper reflection of the Pygmalion myth.

As mentioned earlier, Prospero is modelled on John Dee, the English philosopher,[45] while the name Caliban would appear to be derived from the word cannibal, which Shakespeare probably found in his reading of the French author, Montaigne. A leading idea of Montaigne was that values such as right and wrong and good and bad change with their cultural and historical context.[46]

Caliban is the savage who inherited the island which is the scene of the play. Prospero lands on the island and takes Caliban as his slave. Caliban's wide range of practical knowledge makes him indispensable to Prospero, who has developed the abstract intellect. The basic conflict in *The Tempest* is Prospero's inner struggle to see the figure of Caliban – the vague, the ambiguous, the shadowy, the intuitive – within his own character. When we first meet Caliban, Prospero has already had twelve years of daily confrontation with these shadowy parts of his self.

It could be said therefore that Prospero has been Caliban's apprentice for twelve years. Without Caliban, Prospero would not have survived. His education, and his assimilation of the darker side, is a protracted and arduous struggle. The superior, abstract intellect is reminded of its dependency on the tangible, the practical.[47]

But Prospero also trains Caliban. He teaches him to speak, as Higgins teaches Eliza in *Pygmalion*. Caliban begins to resist. He says that Prospero stole the island where he was 'king'. At the beginning of the play he speaks thus of his captivity:

You taught me language: and my profit on't Is, I know how to curse. The red plague rid you For learning me your language![48]

Prospero gave Caliban a language, and this forced him into a new way of thinking. He finds no associations of his own, and no personal meaning in the words used in his contacts with Prospero. He is in peril of being emptied of his earlier experiences. Together with the Rabelaisian figures of Trinculo the jester and Stefano the drunken butler, Caliban plans to regain the initiative.[49]

Remember,
First to possess his books; for without them
He's but a sot, as I am, nor has not
One spirit to command: they all do hate him
As rootedly as I.
Burn but his books;

Shakespeare is aware of the mechanisms which widen the gap between language and human experience. Crises are confusions of concepts within the system of standards and agreements which allows people to have one and the same perception of reality. What Shakespeare portrays is the crisis which occurs when this kind of consensus ceases to apply, when language and reality no longer coincide.[50]

The emergence of the natural sciences created a new historical context, felt instinctively by the artists. At the time when Galileo was laying foundations of modern science, Shakespeare wrote *The Tempest* (1613)[51] in which he portrays the relationship between the abstract and the practical intellect. The poet foresees the transformation to be brought about by modern science and arms himself against the distortion of the practical intellect which he fears will follow in the wake of this transformation.[52] As we have seen, the movement in this transformation is captured by the authors of the French *Encyclopedia*. There is a connection between my interpretation of *The Tempest* and Rousseau's basic theme: the relationship between two kinds of knowledge. Rousseau thought that if knowledge were divided into two parts, the academic and the general, then the general would be greater than the academic. We have a common, even unconscious knowledge, the breadth and depth of which educated people tend to underestimate, or even, according to Rousseau, prefer to ignore. Theoretical knowledge is not only enriched by experience but without that experience, it tends to create chaos. Rousseau wondered whether the concepts of theoretical science that describe reality can be reconciled with the reality perceived by our senses.[53] This issue will be examined in the next chapter.

Notes

1. GÖRANZON, Bo *Ideologi och systemutveckling. Bidrag till diskussionen om vetenskap, teknik och samhälle* Studentlitteratur, 1978 (andra upplagan), p. 31.
2. EK, Sverker R. *Spelplatsens magi. Alf Sjöbergs regikonst* Nordstedts, 1988.
   As a mature artist, Alf Sjöberg developed a method of working in which he verbalized the idea behind his interpretation. He readily corrected it, and gave more precise and profound explanations of it. Many of his notes bear witness to the way in which he constantly refined his interpretation in the course of producing the play. This form of working in writing became so important for him that he expanded the sketches of his ideas for different productions into full-scale essays. See SJÖBERG, Alf *Teater som besvärjelse* Nordstedts, 1983. See also Laurikainen, K.V. Vetenskapens möjlighet och dess gränser, published in GÖRANZON, Bo (ed) *Den Inre Bilden* Carlssons Bokförlag, 1988, pp. 19–36.
3. SJÖBERG, Alf *Galilei och forskningens frihet* published in Göranzon 1978a, pp. 10–29.
4. Lennart Torstensson, the management consultant at the Swedish Immigration Board, referred to Sjöberg's introductory address in an article called *Utredaretik i statsförvaltningen* published in GÖRANZON 1978a, pp. 52–70.
5. One example was that the National Board of Agriculture was represented by Gunnar Rosquist, a forestry officer who was a colleague of Per-Johan Åge. These two officers were responsible for the development of the forest valuation method. Here, Rosquist had a opportunity to reflect on the conflict among the staff at the national office on the implications of the responsibilities of civil servants. See Göranzon 1978a, pp. 81–83.
6. GÖRANZON, Bo, JONSSON, Inge and MELBERG, Arne (eds.) *Konst och samhällsförändring* Ett samtal i Sigtuna 7–8 October 1977, Report No. 10, SALFO.
7. See GÖRANZON, Bo *Konstens pedagogiska funktion – några exempel* and Melberg, Arne *Estestisk verkan – några problem* published in Göranzon, Jonsson and Melberg, 1978b, pp. 8–36.
8. TURING, A.M. *On Computable Numbers, with an application to the Entscheidungsproblem* London Math. Soc. (2), 42 (1937), pp. 230–265.
9. GÖDEL, Kurt *Collected Works* Vol. 1, publications 1929–1936, Oxford University Press, 1986, p. 136.
10. WHITEMORE, Hugh *Enigmakoden* The Royal Dramatic Theatre, Stockholm, 1988 (Translated into Swedish by Per-Erik Wahlund), p. 26. Published in English as WHITEMORE, Hugh *Breaking the Code* Amber Lane Press, 1987.
11. The play is based HODGES, Alan *Alan Turing, The Enigma of Intelligence* Counterpoint Unwin Paperbacks 1983. A Swedish radio programme on Alan Turing was broadcast in the series *Vetandets värld* on Programme 1 on July 19th, 1988. It was called *Jag vill bygga en hjarna* (I want to build a brain), and presented the ideas contained in the Andrew Hodges' biography of Turing.
12. Whitemore Swedish translation, 1988, p. 26.

13. WEINER, Norbert *Cybernetics, or control and communication in the animal and machine* M.I.T Press and John Wiley & Sons, Inc. 1961 (second edition), p. 23.
14. ROSENBLUTH, Arthur, WEINER, Norbert and BIGELOW, Julien, Behaviour, Purpose and Teleology, *Philosophy of Science*, 10, (1943), pp. 18–24. The interest focuses on a characteristic of teleology or, in other words, 'appropriate behaviour'. This requires a classification of the term *behaviour*. In this classification the word *teleological* is used as a synonym of 'intention controlled by negative feedback'. It means that a stated goal gradually affects the sequence of events with the aim of attaining the goal.

    See also PRINTZ-PÅHLSON, Göran *Turingmaskin* published in the anthology of poetry *Säg, Minns Du Skeppet Refanaut?* Bonniers, 1984, p. 96.
15. WIENER, Norbert *Materia, Maskiner, Människor. Cybernetiken och Samhället* Forum, 1952. Early in his research programme, Wiener established contacts with the prominent social anthropologists Margaret Mead and Gregory Bateson (see Wiener, 1961, p.18). Gergory Bateson published his collected articles on cybernetics in a book *Steps to an Ecology of Mind* Chandler Publishing Company, 1972. See also the chapter headed Den kritiska rationalismen: Karl R. Popper, in JOHANNESSEN, Kjell S. *Tradisjoner og skoler i moderne vitenskapsfilosofi* Sigma Forlag A.S., 1985, pp. 92–115.
16. Wiener, 1961, p. 12.
17. GUILBAUD, G. T. H. *Cybernetik* Aldus/Bonniers, 1962.
18. *Ibid.* p. 96.
19. John McCarthy coined the phrase 'artificial intelligence' as a title for the first research seminar at Dartmouth College, USA, in 1956. Those attending included Minsky, Allen Newell and Herbert Simon.

    See PRATT, Vernon *Thinking Machines: The Evolution of Artificial Intelligence* Basil Blackwell, p. 203, 215.

    See also BOLTER, David *Turing's Man. Western Culture in the Computer Age* Duckworth, 1984, p. 193.
20. TURING, Alan M. *Computing Machinery and Intelligence Mind*, October 1950, pp. 433–460. This work is published in Swedish as *Kan en maskin tänka?* in *Sigma* Vol. 6, Forum, 1960, pp. 2202–2228.
21. HODGES, Andrew Turing's Conception of Intelligence published in Gregory Richard L. and Marstrand, Pauline K. *Creative Intelligences* Francis Pinter Publishers, London 1987, p. 84. The English philosopher A. J. Ayer discusses this paradox of Turing's in the preface to Bolter, 1984, p. XI. The most remarkable thing in this context is, however, Kurt Gödel's refutation of Turing's article Computing Machinery and Intelligence, in which Gödel expresses 'opposition to Turing's mechanistic view of mind', Gödel 1986, p. 25. Wittgenstein made an ironic comment on Turing's article. See GÖRANZON, Bo *Turing's möte med Wittgenstein*, published in Dialoger 5/87 Artificiell Intelligens, p. 43. See also the American philosopher DREYFUS, Hubert L. *What Machines Can't Do: The Limits of Artificial Intelligence* Harper Colophon Books, 1979.
22. Whitemore, Swedish translation, p. 56. Compare with the following quote:

    We may hope that machines will eventually compete with men in all purely intellectual fields. But which are the best ones to start with? Even this is a difficult decision. Many people think that a very abstract activity, like the playing of chess, would be best. It can also be maintained that it is best to provide the machine with the best sense organs that money can buy, and then teach it to understand and speak English. This process could follow the normal teaching of a child. Things would be pointed out, named, etc. Again I do not know what the right answer is, but I think both approaches should be tried.

    (From Turing *Computing Machinery and Intelligence Mind*, October 1950, p. 460.)
23. *Ibid.* p. 438.
24. *Ibid.* p. 439.

    In his article, Turing says that the idea of a digital computer is an old one and refers to Charles Babbage:

    Charles Babbage, Lucasian Professor of Mathematics at Cambridge from 1828 to 1839, planned such a machine, called the Analytical Engine, though it was never completed. Although Babbage had all the essential ideas, his machine was at that time not a very attractive prospect.

    Turing goes on to say that the fact that Babbage's Analytical Engine was to be entirely mechanical will help us to rid ourselves of a superstition:

    Importance is often attached to the fact that modern digital computers are electrical, and that the nervous system also is electrical. Since Babbage's machine was not electrical, and since all digital computers are in a sense equivalent, we see that this use of electricity cannot be of theoretical importance. Of course electricity usually comes in when fast signalling is concerned, so that it is not surprising that we find it in both connections. In the nervous system, chemical phenomena are at least as important as electrical.

25. von NEUMANN, J. *A General and Logical Theory for Automatons* Swedish translation, Sigma, Vol. 6, Forum, 1960.

26. *Ibid.* p. 2194.

I should like to relate this quote to a letter dated 1984–11–30 I received from Lars Löfgren, Professor of Theoretical Automation at the University of Lund.

Dear Brother

I was interested to hear your ideas on the philosophy of computer development earlier this week. As you surely remember I am of the opinion that both Alan Turing and John von Neumann should be included in the circle of enlightened people who have contributed to our scientific understanding of the limitations of formalisation and calculability.

Referring to John von Neumann, I can strongly recommend his posthumously published book The Theory of Selfreproducing Automata edited and completed by Arthur Burks. Urbana and London, University of Illinois Press, 1966.

I was working at the University of Illinois when Arthur Burks was doing the editing work. He went through the main body of material in the form of seminars and one of the things I remember we discussed was the very point which you put forward in support of the idea that John von Neumann had far too optimistic a view of the possibility of making models of cerebral processes. But if you read, for example, pages 46–48 of the book you will see from the context what he meant . . .

May I point out what follows from this from a philosophical point of view, and what does not follow. It certainly follows that anything you can describe in words can also be done with the neuron method . . .

One should bear in mind that this is a transcript of a tape recording of one of John von Neumann's speeches, which Arthur Burks wanted to preserve as much as possible. It is clear that John von Neumann placed a fairly strict interpretation on the phrase 'anything you can describe in words'. He refers quite simply to models in formal languages. Note how he does not just intimate that there are complications when it comes to describing certain phenomena. He even gives an explanation of this with a reference to Gödel (who in his turn refers to Tarski). This understanding of the limitations of language, that there are linguistic phenomena within a formal language which cannot be described within the language itself (which require a higher kind of language), can actually be an argument for a broader language concept . . .

I hope that this may help to bring about a further understanding of John von Neumann's pioneer work in this field.

With kind regards,

Lars

I shall return to this discussion in the next chapter under the heading of Practice – Following a Rule.

27. Descartes designed a tougher version of Turing's test, which we discussed in the previous chapter. The Cartesian test looks like this: before it can be judged to be intelligent, a machine must be capable of language actions and sensible actions independent of the programmer. Descartes arrived at a completely different conclusion to Turing's. The difference between man and an animal – machine is that because he has a language, man is able to develop his thinking and the way he formulates concepts.

28. Bolter, 1984, p. 13.

29. SEARLE, John *Kognitivism och datormetaforer* published in Dialoger magazine, No. 7/8, Artificial Stupidity, 1988.

30. BUTTIMER, Anne *Creativity and Context* Lund's Studies in Geography, Human Geography No. 50, The Royal University of Lund, Department of Geography, 1983, p. 17.

31. SÄLLSTRÖM, Pehr editorial comments in Dialoger magazine, No. 5, Artificiell Intelligens, 1987, p. 4.

32. WEIZENBAUM, Josef *Computer Power and Human Reason: from judgment to calculation* W. H. Freeman and Company, San Francisco, 1967. Weizenbaum's application demonstrates that the *Turing Test* is not a satisfactory one. It can be misleading. This gave Weizenbaum an insight into a fundamental problem: human beings are liable to attribute to new technology,

in this case a diagnostic programme in the field of medical care, more intelligence than it possesses. We lose our distance. We fail to realize what the limitations are. This is a feature of the emergence of a technological culture.

33. *Ibid.* p. 181.
34. Buttimer, pp. 14–15. See also the editorial comments in Dialoger No. 1, Dialogens väsen and Denett, Daniel *The Role of the Computer Metaphor in Understanding the Mind* published in PAGELS, Heinz R. *Computer Culture: the Scientific, Intellectual and Social Impact of the Computer* Annals of the New York Academy of Sciences Volume 426, New York, 1984, p. 274.
35. HILTON, Julian: *Teater och teknologi: Pygmalion och myten om den intelligenta maskinen,* published in Dialoger 6/8, Tyst kunskap.
36. *Ibid.* p. 31.
37. *Ibid.* p. 37.
38. *Ibid.* p. 37.
39. JANIK, Allan *The Role of Literature in the Theory of Knowledge* published in Göranzon and Florin, 1990b. The value of literature in understanding the changes in society and working life was an idea which interested Joseph Weisenbaum (Weizenbaum, 1976, 16, pp. 3).
40. *Ibid.*
41. *Ibid.*
42. An important early inspiration for this orientation was SCHOPENHAUER, Arthur *The World as Will and Representation.* In two volumes, Doven Publications, 1969, and JANIK A. and TOULMIN S. *Wittgenstein's Vienna* Simon & Schuster, New York, 1973.
43. DIDEROT, Denis *Brev till Sophie Valland* translated into Swedish by Olof Nordberg, Atlantis, 1987, letter of September 2nd, 1762, pp. 192–196.
44. See GÖRANZON, Bo *Att se Calibanmetaforen i vår teknologiska kultur* published in Dialoger 9/88 Den andre pp. 17–21.
    Noel Cobb, *Noel Prospero's Island* Coventure Limited, 1984.
    The Royal Dramatic Theatre *Stormen* (The Tempest) Programme, 1968.
    Larsen, Steen *Den arbetande hjärnan. Sammanhanget mellan arbetets organisation och hjärnans funktion,* Prisma, 1982.
45. YATES, Frances *Shakespeare's Last Plays* Routledge & Kegan Paul, 1975, p. 85.
46. de MONTAIGNE, Michel en presentation in *Idehistorisk läsebok,* Vol. 1, Gidlunds, 1982, p. 127.
47. *Ibid.* p. 127., HÅFSTRÖM, Jan *Praktiken i måleriet,* KRIS No. 25/26, 1983.
48. Shakespeare, William *The Tempest.* Compare the meaning of this quote with FROSTENS-SON, Katarina *Språket och den andra,* published in *Dialoger,* 9, 1989.
49. *Ibid.* p. 255.
50. ZERN, Leif *Älskaren och mördaren. Shakespeare och den andra spelplatsen.* Alba, 1984, p. 17.
51. Kline pp. 156–165.
52. See SJÖBERG, Alf *Ögats roll (Troilus and Cressida),* published in SJÖBERG, Alf, 1982, p. 98.
53. DAHL, Eva–Lena *Överideologi och politisk handlingsprogram. En studie i Lockes och Rousseaus tänkande* Acta Universitatis Gothoburgensis, 1980, pp. 179 and 307.

Chapter 5

The Practical Intellect

Is the Computer a Tool?

What is a computer? Computer memory is commonly compared to the human brain and computer 'language' to human language. In fact, there is a profusion of metaphors and parallels about computers; they have at one time or another been compared to steam engines, electricity, the motor car, typewriters, human beings etc.[1]

The question of whether the computer is a tool was the theme of an international conference held in Sigtuna, Sweden, in 1979.[2] Thomas Tempte, the cabinet maker, could not see any marked similarities between a computer and what was commonly understood to be a 'tool' in his trade.[3] Though most of the conference delegates agreed that in one sense or another, the computer is a tool, the philosopher, Tore Nordenstam, the philosopher, pushed the question further: what are the implications of saying that something is simply a tool, and responded:

> . . . as a rule, the introduction of new tools affects the nature of the work itself. Mechanized farming is not only a more effective version of older forms of farming; the tools change the entire orientation of the work. It may be said that the nature of the practice of which the tool is a part changes with changes in the tools used in the practice. As with the practice of the carpenter, the practice of farming is what it is partly because of the availability of certain tools.[4]

The meaning of the word 'computer' is not obvious, nor is the meaning of the word 'tool'. Both these terms are what Allan Janik calls 'essentially contested concepts'. There is no self-evident understanding of what a computer is. Different people place different interpretations on the relationship between man and machine.[5] Allan Janik's point is that these different ideas of what a computer is have a decisive influence on issues related to computer applications.

At the conference in question, Sherry Turkle, a social psychologist at MIT, took up another aspect of computers. She felt that as a tool the computer was slipping through our fingers, and pointed to the need for a more broad-based computer training, derived from philosophy, the history of ideas and anthropology. In her opinion, this training is essential to counterbalance the simplistic stereotype of the computer as 'stupid, doing only exactly what you tell it to do', which is becoming increasingly widespread as many people learn in beginner courses to write simple computer programmes. Large complex computer systems, such as the one in the social insurance sector in the USA, is an example of the opposite extreme. This system had been 'written, rewritten and changed locally by many different programmers and teams of programmers until it has become a patchwork of local modifications, each of which had an inevitable, but often unknown, effect on the total system.' This patchwork is unstable, vulnerable and unpredictable. Sherry Turkle says that this kind of programme is common in public administration and is, in a fundamental way, incomprehensible.

In these social contexts it may be that what is possible 'for the computer' is used to limit the amount of freedom enjoyed by people and organizations. . . . People can begin to forsake their responsibilities in favour of 'the authority of the computer's printout' because that is easier than assuming responsibility within one's work group.[6]

Sherry Turkle's reflections on the effects of computerization in the American social insurance system is an appropriate introduction to the case study of the Swedish social insurance system. This study illustrates two different views of the relationship between man and machine, that is to say, two different views of the nature of computerized administration.

EDP at the Social Insurance Offices – A Case Study

The ALLFA Project

The Social Insurance Offices administer the payment of pensions, sickness and parental benefit payments and other public insurance schemes. This work is governed by a large number of regulations designed to put into effect decisions taken by the Swedish Parliament. The 1960s saw a dramatic increase in the work load of the Social Insurance Offices.[7] In 1967, the Government commissioned the Agency for Administrative Development to examine the possibility of rationalizing the social insurance system by using EDP technology and other means.[8] Computerization began five years later, in 1972. The system consists of a central computer in Sundsvall, and at the time, there were over a thousand terminals at Social Insurance Offices all over the country – these terminals being installed from 1973 to 1976. This computer system with 3200 terminals is one of the largest currently in use in Sweden. Computerized routines include the calculation of sickness benefit payments for normal cases and similar day-to-day tasks in other areas. For example, the calculation and review of pensions is now almost entirely computerized, as is the social welfare payments system.[9] In other words, the computerization of the social insurance system involved the partly automated application of a complicated set of rules. The Government ALLFA enquiry (ALLFA is the Swedish acronym for Social Insurance EDP Systems) was set up in 1977, with the task of examining the introduction of EDP systems in the 70s of the services the offices provided.[10] We were commissioned by the Parliamentary ALLFA enquiry to help 'define and develop the personnel criteria which should be applied in future comparisons and appraisals of different systems for organizing computer operations', to quote the wording of the remit.[11] We designed the research project to include the proposal that the employees' experience of computerization in the 1970s be evaluated and that their views be noted and applied in the next phase of computerization.[12]

The Research Method

This case study, which ran from 1978 to 1981, was carried out in close co-operation with the Insurance Employees' Union. The research method was based largely on an extensive study circle project covering all Social Insurance Offices in the country. We developed basic materials for the study circles, including a summary of available research findings on computerization and skills, one example being the case of the County Agricultural Boards. A great deal of photographic material was also used, intended to encourage reflection.

About 9000 of the 20,000 employees at the insurance offices took part in a total of 900 study circles. Each chapter of the material studied ended with a number of questions for participants to answer. At the end of the study circle, responses were sent to the researchers, allowing an analysis to be made of the employees' ideas and reflections on their work, and the part played by the computer system.[13] The material was evaluated by Ingela Josefson, a language researcher and Peter Gullers, a photographer.[14]

Ingela Josefson observed that the words and terms used to describe computer technology, which also predominate when users of the new technology are trained, are those commonly used among computer suppliers and technicians. This is a world of concepts quite foreign to people who use this technology in their daily work. Of course, the concepts formed in developing computer systems must be precise, but at the same time the specialist language of the computer world cannot give people any idea of what it means to use this technology in a concrete situation. Technical descriptions tend to be anonymous and generate little or no interest when used in training courses. Better results are achieved by using an approach based on the history of ideas to discuss matters like 'what is a computer?' or 'what traditions are followed in the expansion of the social insurance system?'[15]

Peter Gullers, whose photographs on the theme of occupational skills formed part of the study-circle material, described his experience of the study circle as follows:

Passing on knowledge through pictures is not an uncomplicated and problem-free process. But the difficulties involved are almost as poorly-researched as are the possibilities. There are problems related to the fact that the total message of a picture cannot be clearly expressed in words. The contents of the pictures are not exhaustive and translatable. The difficulty in interpreting pictures may lie in the observer being unused to such a process, but it may also be because they put forward genuinely difficult issues and problems with which one cannot deal because one lacks the words and concepts to do so.

Another difficulty is that certain phenomena are not easy to portray. Neither a computer nor the effects of computerization can be made understandable by using words and pictures to describe the machines in simple terms, just as little as hearing can be understood from a description of the ear or a mathematical formula. This understanding requires a portrayal rather than a description. That is the power of the visual arts and poetic language.[16]

Computerized Administration

The ALLFA survey was established to find out what form computerization would take in the Social Insurance Offices from 1985 to 2000.[17] Among other things, a study was made of the possibility of computerizing the handbook which contained the comprehensive set of rules. The attitude of the ALLFA survey was expressed in an interim report published in 1979, which described the advantages of computerized administration.[18] Briefly, the basic ideas were to rationalize and achieve a more uniform treatment of clients:[19]

Work at the processing points to be automated, increasing the speed at which cases can be handled and reducing manpower requirements;

- Conditions to be created for the more consistent application of regulations;
- Fewer training problems in the long term (the work method would be applied in training);
- Maintenance of a higher quality of work, despite staff turnover.

This may be compared with the views submitted by the study circles. The study-circle material contained a question which invited the participants to think about which routine work tasks should be computerized, and the list they produced was a long one. But the computerization of more complex work assignments was rejected completely by all respondents.[20] The following response illustrates the employees' attitude.

A work group's more complicated work assignments cannot be computerized because they require the use of personal judgment. Unfortunately, some of the work group's more complex tasks, such as the reconciliation system for the advance maintenance allowance register, have already been computerized. We cannot give the recipients of these social insurance payments explanations and we cannot influence computerized tasks. Some of the more complicated tasks are often described as fairly simple routine work.[21]

This quote indicates that many of the staff thought that computerization had already gone too far. They considered that more advanced duties i.e. work in which judgment was essential, had already been computerized. It is also apparent that management and staff disagreed on what kind of work should be regarded as more advanced. The following comment from a study circle may be seen as a further development of this line of thinking:

Seen purely on the basis of efficiency, there may be advantages in computerizing processing routines. This kind of arrangement would probably also mean that there would be full compliance with all regulations. However, we consider this kind of arrangement to be completely unacceptable from the viewpoint of both the employees and our clients. A change like this must result in a deterioration of the system that will lead to a lower standard of service, because the insurance office employees can no longer satisfactorily explain the decisions they implement. This must also result in a deterioration in the relationship between the staff and the general public, while the reputation of the Social Insurance Office must also suffer from a general lowering of the standard of its employees as a professional group. Seen from the viewpoint of the general public, some of the documentation issued by the Social Insurance Office may be confusing because computers have a limited capacity to adapt their processing methods to suit a particular case.'[22]

As in the first quote, the staff are thinking about their responsibility towards their clients, specifically their responsibility to explain a decision to people receiving payments under one of the social insurance schemes. The difficulty of providing such explanations is linked to a lower level of knowledge about the insurance schemes which computerization has brought about. Responses to the question about computerizing the handbook pointed to both advantages and drawbacks, as in the following examples:

It would be easier to retrieve the instructions you are looking for. Changes in the regulations could be circulated sooner. This would mean that the handbook would always be reliable, and the administrative processing would therefore also be more reliable than under the present system, where one often has to consult a colleague, who might give you the wrong information.[23]

There would be no advantages. The handbook is always available. The book was produced by human beings to suit the human mind. The photographic memory, leafing through pages by hand, etc., interact to help people remember more.[24]

Employees at the Social Insurance Offices saw themselves therefore both as administrators of the system of rules and as being responsible for its content and application vis-à-vis their clients. This sense of responsibility prevents them from subscribing to the idea of having a uniform way of dealing with all clients, and the staff put forward numerous cases which fall outside the framework of 'normal' situations or cases. Proper assessment of the 'difficult' cases requires judgment based on a well-developed knowledge of the insurance system. The staff meet their professional obligations to their clients through providing a high level of professional knowledge. The serious factor here was that a clear majority of the staff (just on 60%) felt they had lost some of their professional knowledge of the social insurance system since the computer system was introduced.[25]

The final report of the ALLFA committee, published in 1981, questioned computerized case processing at the Social Insurance Offices – see the interim report mentioned above.[26] Among other things, there was talk of introducing direct support for the case officers, but no conclusion had been reached on the form this support might take:

We have also reported that sometime in the future it may be necessary to increase the use of computers as a direct aid in processing cases. The dialogue procedure is one way of using computers to help process cases, giving the case officer access to direct help – when necessary – in the form of information on rules, regulations, etc. stored in the computer. We have not taken a stand on the various possible forms of support for administrative officers – including dialogue procedures – but consider that tests should be completed and evaluated before a view can be expressed.[27]

The report goes on to discuss the dialogue between case officer and computer, and points to the difference between experienced and inexperienced officers:

The dialogue should guide the officer through the procedures step by step. The dialogue must also be designed to permit access to information on current rules, how different computations are made, and the way in which the system performs various checks. The dialogue should be effective for experienced officers inasmuch as it should speed up the process of dealing with the case. It must therefore be possible for the terminal user to avoid being presented with information which he/she already possesses.[28]

This quote also shows that the ALLFA survey left open many questions about the long-term use of computers in the Social Insurance Offices. Two years later – in 1983 – a new committee was appointed, the FAS 90 Committee, with the task of continuing the work of the ALLFA Committee. This committee suggested that expert systems appeared to be the solution to the problem of computerizing the more specialized aspects of social insurance work, the expert system guiding the officer through the case in question:

A special category of decision-supporting systems is the so-called expert system (knowledge-based systems for aiding decisions). Here, the system of rules and other 'expert knowledge' on the method of dealing with social insurance cases are to be stored in a database. The system guides the officer through the process of dealing with a case. The officer keys into the system certain data on the case, and the system can respond by producing, relevant to the case under consideration. A dialogue should be possible between the case officer and the system, in which the system can request further data on the case, and the officer should be able to request more detailed explanations of the conclusions produced by the system.[29]

With or without an expert system, however, the basic aims remain the same: rationalization and the standardization of the processing of social insurance payments.

When the committee produced the outline of a prototype early in 1986 it was stated that most case officers should be able to make decisions with the help of the expert system without consulting their colleagues. The background to this statement was that in the course of their visits to a Social Insurance Office in Stockholm, the researchers had observed that case officers spent a great deal of their time talking to one another. The researchers therefore thought that limiting communication among case officers was one way of rendering the system more effective.[30]

The committee's attitude to work and professional knowledge is also expressed in a report on a visit to the UNITED KINGDOM in 1986, where trial projects were in progress on decision support systems in social insurance as part of the national investment in information technology, the so-called ALVEY programme. The report on this fact-finding trip described the way the British intended to use expert systems in training. 'Training is supported through the development of a prototype that is an aid to training intuitive or tacit knowledge. This is a skill that one acquires by working with these questions, thus acquiring common sense which can be applied in that context.'[31]

The official view stated in the ALLFA report and the subsequent investigation disagree with the staff's comments on their work and computerization. Moreover, just as the authorities had clung obstinately to their basic ideas over the years, the staff too defended their view that it is vital to have a sound knowledge of the insurance system and that as social insurance office staff they have a professional responsibility to their clients, the people covered by the insurance system. There is further evidence of this in a quote from a conversation in 1989 between Karl-Olov Arnstberg, the ethnologist, and a Social Insurance Office employee which took place about ten years after the major study circle project was run:

Is this a job which a fairly intelligent computer could take over?

No, I don't think so. I think that we have to be here ourselves. I hope so, for the sake of our clients. A really horrifying view of the future we have would be two case officers sitting in an office when someone comes in to talk about parental benefit. As a case officer you should be able to press a button and display the administrative routine on the computer screen and read it off. You would not have to know about it yourself, you would simply have to read it off the screen. And that is the way 'they' want it to be.

In his study Arnstberg touches on some aspects of professional knowledge which are related to expert systems, one being the question of collective competence. Arnstberg observed that there was a very high level of collective competence at the office he studied, which meant that the case officers were constantly trading knowledge with each other:

Competence essentially becomes something we all carry in our heads. So when someone leaves the office it is not simply a matter of a person leaving, but some professional competence is lost as well.[33]

This comment provides an explanation for the amount of time that the case officers spent talking to one another. The point is raised again in the following joint statement on computerization and future professional knowledge made by employees from different Social Insurance Offices:

We believe it is very important to think about which work assignments we want to automate. It is not certain that everything which can be computerized should be computerized. . . . A local office usually has a considerable level of collective professional knowledge which should be valued very highly because it raises the overall level of knowledge in that office. . . . We feel it may be necessary to sacrifice some of the apparently endless possibilities of technology in favour of a manual, perhaps slower but 'more human' way of working, in order to prevent contacts with colleagues becoming less frequent and dependency on the computer as a working partner rather than on our colleagues.[34]

The following is another extract from Arnstberg's study on expert systems dealing with case officers' professional knowledge and specialist language:

The knowledge possessed by case officers may be described as being directly linked to their work tasks. . . . The central issue is how the matter is to be processed. Their specialized professional language becomes detailed and precise. When talking about their jobs, the insurance office secretaries' language is detailed and subtle. They use in their work an elaborate language which they do not use in the same way when they talk about their work. This becomes clear to me when I compare the language I have listened to with the language used in the interviews. The language I listened to is considerably more precise, to the point where it may be difficult to transfer professional competence between different kinds of work situation.[35]

All this suggests a problem area in the development of expert systems. Given the fact that the expert system requires employees to provide exhaustive descriptions of their work, how can the 'knowledge engineers' overcome the employees' genuine inability to describe their professional knowledge.[36]

In his study, Arnstberg discusses both computerization and how an ongoing process of reorganization affects the professional knowledge of the executive and administrative staff. He discusses the extent to which case officers feel certain of their judgment.

Their criticism of the latest reorganization is that it appears to create a work situation where it is not possible to achieve certainty. . . . Only parts of the reorganization affected 'my office'. There were still 'stars' who knew everything about specific areas, but these people were afraid that they would not be able to maintain their competence in the new system. Their competence became broader but less profound. In other words, they had less certainty.[37]

There is a parallel here between reorganization and computerization. The fall in the level of knowledge on insurance matters noted after computerization had taken place was precisely a question of employees being less certain about their own judgments.

The Inner Weather Picture

The case study from the Social Insurance Offices had an important effect on the future orientation of our work. It confirmed that professional knowledge was an important theme which needed to be researched on an epistemological basis. Professional knowledge is not an issue which is subsumed to working environment or employment policy considerations and, first and foremost, the relationship between professional knowledge and the use of computers is something quite separate from systems development and design.[38]

What the Social Insurance Office staff were talking about comes under the heading of the practical intellect, and the remainder of this chapter is devoted to an examination of this phenomenon. We shall make a closer examination of the nature of professional knowledge, and in doing so we shall also examine the criticism of people's excessive faith in the abstract intellect.

The first piece of the puzzle is taken from a study of meteorologists and their professional knowledge, in which Maja-Lisa Perby introduces the inner picture as a metaphor for professional knowledge.[39] We are reminded of Thomas Tempte's use of the same metaphor in his reflection on the reconstruction of Tutankhamun's chair, or in the context of the Social Insurance Offices the use of 'personal judgment.'[40]

In her study Maja-Lisa Perby asks: how do meteorologists make their judgments in a real-life situation? how do they use their own knowledge and the wealth of information and technical aids available in order to arrive at a weather forecast? Her answer is that the meteorologist builds an inner picture of the weather for the day. It is this inner weather picture which allows him to judge the way the weather picture will develop and make a forecast.[41]

A layman cannot really form an idea of the inner weather picture. It is the weather experienced both by the senses and through the values of abstract parameters. It is both a theoretical interpretation and one based on experience; it is both fact and assumption. It takes time for the meteorologist to become familiar with the day's weather. He gradually builds up his inner weather picture by virtue of his training and experience and by using certain routines to acquire various kinds of information.

The inner weather picture is at once individual and collective. Every meteorologist builds up his own picture of the weather, but in his work he follows a professional tradition which governs the way information about the weather is used. One consequence is that colleagues can exchange ideas on probable changes in the weather picture, and note that they are in agreement although they cannot even explain why they believe the weather will develop in a certain way. This is illustrated in the following quote:

It is difficult to say why you believe that there will be certain changes in the weather. You have an overall picture of the weather which you use as a starting point.[42]

The metaphor of 'the inner picture' is a shift in perspective: the content of the metaphor is different in the man-machine relationship than in the generally accepted interpretation; one approaches questions related to information and technical systems from another angle. For example, the most important aspect of information becomes the extent to which the information can be fused with the inner picture. To take an example from the meteorological study, Maja-Lisa Perby notes that satellite and radar pictures are technical aids which the meteorologists themselves require; they provide information which fills the gaps in the inner weather picture. At the same time, satellite and radar pictures were introduced into a work situation where many technical aids compete for attention. Meteorologists have no time to assimilate all the information available. They must select and eliminate; and this is a process which affects the inner weather picture.[43]

Using the inner weather picture, Maja-Lisa Perby can interpret several complications which technical innovation has brought to meteorologists, even complications which initially appear rather odd. Because numerical forecasts, their most sophisticated aid, have for various reasons proved demonstrably difficult to integrate into the inner weather picture, meteorologists are unable to make a skilled assessment of them.[44]

The inner picture also draws attention to the fact that many different circumstances in the work and at the work place may affect the possibility of making good judgments. 'Long thoughts', for example, require time, that is time to digest information and let it fall into place. In other words it may be more important to have a cup of coffee and let one's thoughts about one's work roam free rather than to turn to yet another technical aid. As Diderot says, rhythm in work requires a pause for reflection, and this pause may be essential if the quality of the work is to be maintained.[45]

Every occupational group and every work situation is unique, but Maja-Lisa Perby's study draws attention to the need to respect what constitutes the core of an occupational group's ability to make judgments. Thus in the context of forest valuations, it is important to know: what conditions in the forest rangers' work help – or hinder – them to build up a well-founded inner picture of a forest property?

Practice: Rule-following

The next step in our analysis of the practical intellect and working knowledge is to examine some of the most important ideas of the philosopher, Ludwig Wittgenstein. This will also help to clarify the inconsistencies and conflicts we met earlier. In his later philosophical thinking Ludwig Wittgenstein concentrated on the kind of problem that lay behind the conflict among staff at national level in the County Agricultural Boards, i.e. differences in perceptions of reality which stemmed from differences in experience.

The multiplicity of activities, or differences in attitude, captured Wittgenstein's interest. In *On Certainty* he expresses it like this: 'Compare the meaning of a word with the function of a clerk. And different meanings with different functions.'[46] He focuses attention on the functions in an organization because different functions involve, and are based on, different experiences of reality.

What does following a rule mean? In asking this question Wittgenstein leads us into a discussion of human actions or practices and, not least, of certainty in action. In his mathematics lectures from 1942 to 1944, Wittgenstein discusses calculating machines. In *Remarks on the Foundations of Mathematics* he says:

Imagine that calculating machines occurred in nature but that people could not pierce their cases. And now suppose that these people used these appliances. . . . Thus for example they make predictions with the aid of calculating machines, but for them manipulating these queer objects is experimenting. These people lack concepts which we have; but what takes their place?[47]

What does Wittgenstein mean by the word 'concepts' in the last sentence? There is a vital difference here from what Turing sees as constituting a concept. Wittgenstein says something quite different. He does not focus on the rule itself, but on following a rule. Let us examine this distinction in more detail.[48]

Wittgenstein sees a concept as a set of activities following a rule. The meaning of the concept is determined by its use. It is our traditional usage or practice (Wittgenstein uses this term) that best illustrates the nature of our fundamental understanding of something.[49] In other words, the rule is built into the action. This means that the essence of a practice cannot be expressed in a formal description.

The knowledge contained within a practice cannot be passed on to other people directly. Wittgenstein says that we are taught a practice through examples, from models and by training. It can be passed on to some people by analogies and concrete examples. At the same time the individual must strive to gain a deeper insight into the practice and its working knowledge.

The teacher should illustrate by examples. There are 'good' examples which lead our thoughts in the 'right' direction and which refresh our minds, and there are examples which make it impossible to understand the sense of a practice.[50] Wittgenstein says

. . . but if a person does not yet have the concepts, I should teach him to use the words by means of examples and by practice.[51] . . . If language is to be a means of communication there must be agreement not only in definitions but also (strange as this may sound) in judgments.[52]

One result of Wittgenstein's view is that special emphasis is placed on learning from practice. There is a parallel here with Frances Yates' views on apprenticeship training and the art of memory (see Chapter 3). One does not learn a technique, one learns to make proper judgments by gaining personal experience.

Another illustration of Wittgenstein's distinction between a rule and following a rule may be found in Ingela Josefson's description of how Social Insurance Office staff build up their professional knowledge

> . . . The staff's professional knowledge was a combination of knowledge of the contents of the relevant legislation (i.e., the rules) and the practical experience they gained when in their contacts with their clients they met complications, which applying the law so often involves. The latter relates to following a rule.[53]

Wittgenstein's emphasis on action, practice, also explains the disagreement between the state authority and Social Insurance Office employees concerning the line to be drawn between routine and qualified work tasks. Taking part in a practice allows one to benefit from the experience of others. Previous experience and problem solving, so-called sediment,[54] is turned into a process for following rules which forms the basis of the practice that we are being taught.[55] 'If our certainty is grounded on experience, then naturally it is from past experience – not just my experience but other people's – that I gain knowledge.'[56]

One consequence of Wittgenstein's view is that acquiring skills and expertise requires work to be organized in a way that allows social contacts among employees to be maintained. This is essential if knowledge is to pass from master and expert to trainee and competent employee.[57]

Evident in Wittgenstein's reasoning is an imperative aspect to following rules in a practice if one is to go beyond these rules and bring a creative attitude to new situations. The following quotation from Thomas Tempte's example of boat building and apprenticeship illustrates this point but it could just as well apply to forest rangers, Social Insurance Office staff, meteorologists, nurses or teachers:

> Breaking set rules is not primarily an opposition to traditional authority but a question of growth, of expanding beyond the rules. It is reflection, not defiance, which is needed . . . and in a given situation one must arrive at a realization and sum up in one's mind: there is here a turning point after which one cannot go back, only forwards. If you have been working long enough you have learned to trust your intuition, your craft skills. . . . Stability and continuity. The changes are material ones, but the act of performing the work is the same.[58]

There are people who maintain in all seriousness that because computer technology is used today in many different kinds of work it is easier to move from one field to another, providing one has mastered the technology.[59] This is the opposite to emphasizing the aspects of a practice which are specific to the work area, as can be seen from Wittgenstein's reasoning. Analogies and examples must be taken from within a practice, and they cannot be transferred from one area to another. For example, one cannot transfer from error location on aircraft to error location in a computer programme for forest evaluation as they are two quite different trouble-shooting practices.[60]

Thus the deciding factor is the ability to see the differences between different activities. On the other hand, it is just as important to perceive the continuity in mastering error location, for example, when computerizing an operation.

Certainty in Action

No matter how sophisticated the forecasts produced by computer systems are, they do not make the meteorologist feel certain in his judgment of the weather. One of the findings of the study of meteorologists was that the latter must be able to use the computer's forecasts together with their own forecasts of the weather picture.

The question of certainty when acting in a practice lies at the core of the practical intellect. In the two case studies, from the County Agricultural Boards and the Social Insurance Offices, the employees felt that they were less certain in their judgments than they had been before computers were introduced. A reorganization of the Social Insurance Offices introducing broader work tasks resulted in a fall in the level of certainty in work.

Wittgenstein discusses how our certainty in acting in a practice is affected by the introduction of more technology. In his *Remarks on the Foundation of Mathematics* he writes:

We may trust 'mechanical' means of calculating, of counting more than our memories. Why? – Need it be like this? I may have miscounted, but the machine, once constructed by us in such-and-such a way, cannot have miscounted. Must I adopt this point of view? – 'experience has taught us that calculating by machine is more trustworthy than by memory. It has taught us that our life goes smoother when we calculate with machines.' But must smoothness necessarily be our ideal (must it be our ideal to have everything wrapped in cellophane)?

Might I not even trust memory and not trust the machine? And might I not mistrust the experience which 'give me the illusion' that the machine is more trustworthy?[61]

He continues: 'When I know how to act in every particular case, this means that I can act without hesitation, it is self-evident to me. I say 'of course'. I can give no reason.'[62] Wittgenstein then asks where this certainty of action comes from?

Is it not enough to observe there is certainty in action? Why should I look for a basis for certainty? If you understand 'what it means to follow a rule', you must already have the competence to follow a rule.[63]

We may conclude therefore that if people who are active in a practice begin to express doubt about their certainty in action, there is reason to be cautious. This may be expressed even more strongly: as soon as a skilled person begins to question his certainty, there is reason to be alert as to what is happening in terms of professional knowledge.[64]

Three Categories of Knowledge in a Practice

Following a practice means applying what we may call practical knowledge, knowledge based on experiences gained from being active in a practice.[65] At the same time we learn a great deal from the examples we are given by others who are working within the practice. This latter kind of knowledge, acquired from learning a practice by seeing or examining examples of tradition in the work, we may call the knowledge of familiarity.[66] It is from this aggregate experience, and not from first-hand experience alone, that we build up competence. Interaction among people in the same professional group is crucial.

The part of a professional tradition expressed in general traditions, theories, methods and regulations and capable of being assimilated through a theoretical study of an activity, we may call propositional knowledge.[67]

In short, knowledge may be categorized as follows:

1. Propositional or theoretical knowledge
2. Skill, or practical knowledge
3. Knowledge of familiarity

There is a clear tendency to attach too much importance to theoretical knowledge at the expense of practical knowledge, while the knowledge of familiarity is often completely ignored in discussions of the nature of knowledge.

The relationship between the three kinds of knowledge in a practice may be described as follows: we interpret theories, methods and regulations through the familiarity and practical knowledge we gain from being active in a practice. In the dialogue among people involved in a practice, there is some friction between the different perceptions people have based on their different experiences (different examples of familiarity and practical skills). Being a member of a practice while at the same time acquiring greater competence requires a continuous dialogue. Being professional means extending one's perspective to encompass a broader overview of one's own skills. According to this argument if we remove all the practical knowledge and knowledge of familiarity from an activity, we will also empty it of propositional knowledge.[68]

The interdependence of different kinds of knowledge, is central to Wittgenstein's philosophical work. It is closely linked to his perception of the unsayable as the cardinal problem of philosophy.[69] In *Culture and Value* he says: 'The unsayable (what appears to me to be mysterious and which I cannot bring myself to express) forms perhaps the background which gives meaning to what I could say.'[70] Here, Wittgenstein draws our attention to two kinds of knowledge – explicit and tacit knowledge. However, the latter type of knowledge has been much distorted.[71] The philosopher, Kjell S. Johannessen, who divided knowledge in a practice into three categories, has criticized the usage of the term tacit knowledge and attempted to make it usable once again.[72]

Transferring a Working Culture

There are two sides to knowledge, the technical and the normative. This important distinction, not least in an educational context, was introduced by Jan-Erik Degerblad in a thesis on the scientific theory of planning.[73] Degerblad placed both propositional knowledge and the knowledge of experience (the mastering of the tools in an activity) on the theoretical side of knowledge. He places on the normative side the kind of knowledge which contains good or bad examples/models, the kind of knowledge which is expressed by our 'knowing that'. This kind of normative knowledge then forms the basis of the kind of knowledge that gives us the ability to make judgments in the activity in question. The normative side may be illustrated by the following quote from Wittgenstein's *On Certainty*. 'Rules do not establish a practice, one must also have examples. Our rules leave back doors open and a practice must speak for itself'.[74] Wittgenstein also said that as we learn rules, judgments are made, and their context gives us more judgment. Overall judgment is then made credible.[75]

The normative aspect of an activity means that we cannot simply talk of transferring practical knowledge when we introduce someone to an activity. There is a transfer of both technical and normative aspects – the transfer of a working culture.[76]

Socrates as a Knowledge Engineer

The American philosopher, Hubert Dreyfus, and his brother, Stuart Dreyfus, the mathematician, referred in their book Mind over Machine to one of Plato's dialogues, *Euthyphro*.[77] The authors used this dialogue to discuss the relationship between experts and knowledge engineers – the latter being the term used for people who tap the expert's knowledge and transfer it to an expert system. In the above mentioned dialogue Socrates wishes to clarify how dedication can be characterized. As a religious prophet, Euthyphro gives examples from his work. In this dialogue Socrates thinks that Euthyphro should formulate the rules which characterize a dedicated action. This is the attitude of a knowledge engineer.

Euthyphro claims, however, that he can assess what a dedicated action is – but cannot describe the rules which control assessment. Plato says that experts, in this case a religious prophet, have once known the rules they use but then forget them. The task of the philosopher is to help people remember the principles which control their actions.[78]

Transferring Ethics in a Working Culture – a Socratic Task

The physician, Martin Fahlén, has given an interpretation of the Socratic method in a study in which he uses Socrates as a model for the apprenticeship system:

Every task requires a certain level of competence, and when two doctors are working together, their levels of competence should be different, for example, one an experienced doctor, the other with less experience. The experienced doctor becomes a mentor. A kind of 'Socratic' process ensues. A specialist could do the same work on his own, but the educational factor would be lost.[79]

It is generally thought that Plato/Socrates placed a high value on the abstract intellect and belittled practical work, arguing that people with practical skills cannot compose a text describing how they carry out their work. The carpenter cannot describe how he makes a chair, but because the philosopher can discuss the concept of a chair and its use, he is superior to the carpenter who made the chair.[80]

A point is made in Plato's dialogue, *Georgias*, which is of interest to our discussion. There, Socrates denies holding a disparaging view of practical skills. Rather, he simply points to an incontestible fact – that practical skills can be used for both good and evil. Insight is lacking, an understanding of what one is doing and what purpose it will serve. This intellectual dimension, the 'inner dialogue', is essential before a skill can be called 'art'.[81]

A doctor well-versed in the art of medicine will prescribe a course of action which may cause discomfort but will lead to a lasting improvement in the condition. A doctor who only meets the patient's immediate need for relief from discomfort is a bad doctor.

Here, can see two levels. At one level, there is the relationship between language and action – the extent to which we can describe our actions – a phenomenon discussed by Dreyfus in Euthyphro when he introduces the idea of the knowledge engineer. The second level, the need for insight and consideration, postulates the ethical dimension depicted in Georgias and which Martin Fahlén probably had in mind when he advocated the transfer of a working culture from an older, more experienced physician to a younger one.

Expertise

Moving from the dialogue of the Greeks to that of contemporary experts, Per Svensson, has been compelled to reflect on what works in practice. He considers that users learn to find errors in the computer programme in a manner that defies explanation. This in its turn has consequences for the systems expert's work and his professional skills:

In the routines at the Department of Agriculture for valuing forests using EDP, error location and the correction of input data is one of the most important jobs. Programmes have been written that search through input data and report any errors, and they are controlled by given rules that are part of the programme. It is impossible to write programmes that locate and make a perfectly clear report on every kind of error. The input data varies far too much for this to be a practical possibility. Instead, users must learn by experience. Having worked with this application for a long time, the speed with which most experienced users now locate these errors is incomprehensible to new employees. When asked how they do it, they answer: 'I see that it is an error'. One explanation of why experienced users recognize errors when inexperienced users do not discover them is that their experience contains memories from earlier, similar cases, even if one cannot with certainty report when they occurred. This is a form of knowledge that is extremely difficult to document, but which nonetheless exists and works in a practice.

Attempts have been made to document this particular kind of work. The experience gained from these attempts is daunting. The result of the documentation was a very comprehensive catalogue of every imaginable error, how they were reported by the programme and the action to be taken. For new users, this catalogue was both frightening and of little use, while experienced users worked quicker and more surely if they trusted their own experience and did not use the error location catalogue. Experience cannot always be documented in a usable way.[82]

This quotation brings us to a general point related to the way computer operators learn to use the system. Once again this concerns the practical intellect. It is extremely difficult to produce comprehensive and at the same time, readable documentation for computers systems of any size, and even more difficult to learn the system simply by reading this documentation. Of course documentation is needed to allow new staff to familiarize themselves with the system, but documentation must assume that practical experience will be gained from actually working on the system.

Apprenticeship shop training. Transferring a working culture

Competent workers can follow rules and work well, but their competence lacks direction when normal conditions no longer obtain. When problems occur, a competent worker needs the help of an expert, just as an athlete or an artist needs a trainer or instructor to consult on technique. The reason for this is that in one sense competence is an uncritical understanding in practice, something which requires expertise in situations where normal conditions no longer apply, i.e., expertise may be equated with criticism or a critical analysis of exactly what has gone wrong. Once the expert has got to the bottom of the problem the competent worker can then begin to resolve it.[83]

Supervision of paper production, Värnamo Well, 1969. Photo: Peter Gullers

Another effect of using EDP is that when a work task is computerized one loses a larger part of professional knowledge than that which is specifically related to the computerized task. There is always a connection with other similar routines. If one a single task is computerized, connections with other tasks are also broken. This means that one also loses some knowledge about parts of the routines which are still carried out manually.

EDP is mainly used as an instrument of rationalization. There is no time for reflection and assessment, or for necessary contacts with the clients.

(Quote from Håkan Svensson of the Malmö Social Insurance Office at the Conference on Education-Work-Technology run by the Nordic Council of Ministers and the Arbetslivcentrum, 30 September–2 October 1985

Is the familiarity with reality obtained from manual contact – from fingers, hands and eyes – of any value? If we believe that something is lost when calculations are no longer made manually, can that loss be replaced by something else? The long-term effects described here are really long term, they do not appear before the next generation move in to the expert group. This is what is happening now, because the age difference among the experts (forest rangers) means that a large number of will retire in the next few years and will take with them the really deep knowledge of how forest valuation calculations are made.

A new system of forest valuation is being designed, harnessing the speed of the computer, which may lead us into a completely new situation: either one makes an expert system with calculations which are so difficult that they have to be done on a computer, or a system is constructed in co-operation with the valuator, using the computer as an optional aid. In this case, the valuator can make calculations without using a computer.

This is a difficult path, and one which requires a serious discussion at a fundamental level of the computer as a tool or the computer as a controlling machine.

(Quote from Per Svensson, systems expert, the National Board of Agriculture at the conference on Education-Work-Technology run by the Nordic Council of Ministers and the Centre for Working Life, 30th September–2nd October 1985).

Notes

1. By the same token there are numerous analogies of what man is in the discussion of technological change: man as a clockwork mechanism, an ant, a piano etc.
   See, for example, SIMON, Herbert *The Science of the Artificial* MIT Press, paperback edition 1970, in which he compares man with the ant.
2. SUNDIN, Bo (Ed.) *Is the Computer a Tool?* Almqvist & Wiksell, Stockholm, 1980.
3. *Ibid.* pp. 65–69.
4. *Ibid.* pp. 57ff.
5. *Ibid.* pp. 70–80.
6. *Ibid.* p. 81.
7. The Social Insurance Employees' and Insurance Agents' Union: *ADB inom försäkringskassorna.* (A programme of action for the insurance employees' union) 1980b.
8. *Ibid.* p. 7.
9. *Ibid.* p. 8.
10. Göranzon, Bo: (Ed.) *Datorutvecklingens Filosofi. Tyst kunskap och ny teknik* Carlssons 1983, pp. 194–198. An appendix in this book contains the committee's remit for the ALLFA research project.
11. *Ibid.* p. 193 (letter from the Ministry of Health and Social Affairs, 1977-11-21 to the newly founded Centre for Working Life).
12. This case study, which ran from 1978 to 1981, was carried out in close co-operation with the Insurance Employees' Union. The research method was based largely on an extensive study circle project covering all Social Insurance Offices in the country. We developed basic materials for the study circles, including a summary of available research findings on computerization and skills, one example being the case of the County Agricultural Boards. A great deal of photographic material was also used, intended to encourage reflection.
    About 9000 of the 20,000 employees at the insurance offices took part in a total of 900 study circles. Each chapter of the material studied ended with a number of questions for participants to answer. At the end of the study circle, responses were sent to the researchers, allowing an analysis to be made of the employees' ideas and reflections on their work, and the part played by the computer system. (GÖRANZON, Bo *et al.*: *Job Design and Automation in Sweden. Skills and Computerization*, Report 36, The Centre for Working Life, 1982, pp. 57–101.)
13. The Social Insurance Employees' and Insurance Agents' Union *DATORN: Studie och informationsmaterial om datoranvändning i försäkringskassorna.*
    At the end of the study circle course, the responses were sent in, processed and collated. The responses, which were all of the qualitative type, were grouped under main headings. The size of the groups varied according to the problem area. Because a study circle could send in a response which was classified under more than one heading, the totals for some responses were over 100%.
    The entire study circle course project could be seen as a problem inventory of a computer system which had been in use for seven years and which was one of the most complicated computer systems in use in Sweden at the time. The material was evaluated by Ingela Josefson (a language researcher) and Peter Gullers (a photographer).
14. JOSEFSON, Ingela and GULLERS, Peter *Begripa och förstå. Forskning om metoder att örmedla resultat i arbetslivsforskningen.* The Swedish Centre for Working Life, Report No. 8, 1983, and JOSEFSON, Ingela (Ed.), *Språk och erfarenhet* Carlssons, 1985.
15. Josefson, 1985, pp. 160–166.
16. *Ibid.* p. 189.

17. Göranzon, 1983, pp. 160–166.
18. The ALLFA project *ADB inom den allmänna försäkringen – på 1980-talet och därefter*, interim report DS 1979: 4.
19. Göranzon, 1983, pp. 215ff.
20. The Social Insurance Employees' and Insurance Agents' Union: *DATORN: Studie och informationsmaterial om datoranvändning i försäkringskassorna.*
    When asked to describe what routine tasks and more complex tasks can reasonably be computerized in the group's area of work, the replies fell into two main categories:
    1. No computerization of more routine tasks: 227 circles = 32%
    2. Examples given of the computerization of routine tasks: 460 circles = 65%
    The responses to the question: 'describe the advantages and disadvantages of transferring putting the handbooks into the computer' were grouped into three main categories, with the advantages and disadvantages listed for each category.
    Advantages:
    1. No advantages: 202 circles = 29%
    2. A complete and up-to-date handbook and more rapid processing: 354 circles = 50%
    3. A uniform service to the recipients of insurance payments: 54 circles = 8%
    Disadvantages:
    4. Deterioration of social insurance office employees' professional knowledge: 406 circles = 58%
    5. Computer breakdowns have negative effects: 279 circles = 40%
    6. Other views (central control, impossible to make personal notes, worsened social contacts in the work group, more computer terminals): 170 circles = 24%
    The responses to the question 'Give examples of the kinds of case in which the computer system makes it more difficult to take into consideration the client's particular situation', fell into two main groups.
    1. Cases involving judgment and/or language problems: 446 circles = 63%
    2. Caused by shortcomings in the technical/administrative routines: 148 circles = 21%.
21. The Social Insurance Employees' and Insurance Agents' Union, 1980, p. 11.
22. *Ibid.* p. 23.
23. *Ibid.* p. 27.
24. *Ibid.* p. 28.
25. *Ibid.* p. 25, see Note 20.
26. The ALLFA project Social örsäkringens datorer SOU 1981:24.
    The research project carried out in connection with the computerization of the Social Insurance Offices included an historical study which outlines two traditions in the field of social insurance. This study is published in Thomas Fürth: *Organisationssyn och teknikval inom offentlig administration 1930–1980. Exemplet social örsäkringen* in Göranzon 1983, pp. 119–137. This study gave the employees a perspective on the current situation and was a productive aspect of a discussion on the development of computers up to the year 2000. The historical description put the administrative authority into question.
    In an official comment on the research report, the authority says that there is surely no disagreement with regard to the goals 'but on the other hand, there may be differences of opinion on which methods are appropriate for reaching the goals'. In a separate study, *Värderinger och paradigm vid datasystemutveckling: exemplet ALLFA utredningen* The Swedish Centre for Working Life, 1980: 27, the philosopher, Tore Nordenstam, discussed the implications of separating goals from means. The crucial point of Nordenstam's study is to demonstrate how the use of a certain set of concepts can express given interests. Values and interests become part of the language.
27. *Ibid.* p. 20.
28. *Ibid.* section 6.5.4 of 'Dialog'.
29. The National Social Insurance Board, Request for appropriation for the development of a decision support system in the field of social insurance, 1985-10-6, p. 2.
30. The Agency for Administrative Development. A sub-project on the information system FF. (A report of a series of interviews) PM 1986-06-17, and The National Social Insurance Board Research assignment on the long-term orientation of ADP work in the social insurance system, etc Draft Report 1986-09-xx.
31. KLOCKARE, Barbara, Report from a visit to ALVEY Demonstrator Project, Agency for Administrative Development, 1986-06-11.
    This report states that: 'it will take 5–6 years for experts systems to begin to give results and be applied in a sensible way and become commercially viable'. Compare WINOGRAD, Terry

and FLORES, Fernando *Understanding Computers and Cognition. A new foundation of design* Ablex Publishing Company, 1986, where the authors see the computer as a 'coacher' to develop 'social agreements' in work. The computer will give the users the opportunity to develop their social competence.

This kind of infringement of traditional boundaries was noted by Aristotle in *The Nicomachean Ethics* OUP 1984, where he maintains that a sign of an educated man is that in each subject he only attempts to achieve the degree of precision which the nature of the subject permits. This may result in a false description of reality if one uses more exact terms than the subject allows. This theme is discussed in GÖRANZON, Bo *Bildning vid systemutveckling. En förståelse av den mänskliga dialogens karaktär*, published in AHLIN, Jan (Ed.): *Konsekvenser för industri- och arbetsmiljöplanering ny informationsteknologi*, Project Report No. 3, Department of Architecture of the Stockholm Institute of Technology, 1985, p. 101–120. This paper is a comment on *Systemutmutveckling. Presentation av fyra oliksa sysnsätt*, The Development Programmme for New Technology, Work Organization and the Working Environment, The Swedisk Work Environment Fund, 1984.

32. ARNSTBERG, Karl-Ov *Tjejerna på kassan. En etnologisk studie av yrkeskompetensen på ett försäkringskassekontor*, Work report, The Centre for Working Life, 1989, p. 61.
33. *Ibid.* p. 46.
34. ECKERBOM, Gertie (co-ordinator) *De framtida yrkeskunskaperna – några synpunkter från förs/F6 kskontoren*, published in Göranzon and Bergström (Ed.): *De framtida yrkeskunskaperna. En arbetsbok ör reflektion* The National Social Insurance Board, 1990. This document is the product of a series of seminars run in connection with the FAS 90 research programme in the Spring of 1989. This programme planned and carried out this series of seminars.
35. Arnstberg 1989, p. 35.
36. IVA *Important Technological Trends. Artificial Intelligence and Computer Science*, IVA report 246, Stockholm 1983, pp. 31–32. The report states that there is 'a desperate need to find methods' to capture man's diversity and flexibility in problem-solving. See also HART, Anna: *Knowledge Acquisition for Expert Systems*, published in Göranzon and Josefson: *Knowledge, Skill and Artificial Intelligence* Springer Verlag, London, 1988, pp. 103–111.
37. Arnstberg, 1989, p. 56.
38. This distinction affects the design of research programmes on themes such as people, technology, society. In the last chapter of this book, I discuss some reflections on this subject. See also the programme for the conference on *Culture, Language and Artificial Intelligence* Stockholm, May 30th–June 2nd 1988, and the two books from the conference: Göranzon and Florin 1990a and Göranzon and Florin 1990b. Concomitant with that international Conference, a national postgraduate course on the theme of Computers and Knowledge was run under the auspices of SALFO/FRN. Documentation from this summer academy is published in Göranzon *Datorer och kunskaper* (work report) 1988. See also the magazine *Dialoger* 10, Datorer och kunskaper. According to Thomas Kuhn *The Structure of Scientific Revolutions* 2nd edition, Chicago 1969, a new paradigm is built up through long-term work in the form of conferences, course literature, course development, magazines etc . . . We have built up this network of activities under the heading of *Education, Work, Technology* over the last 10 years. See FLORIN, Magnus *Kunskaper, arbete, utbildning – och artificiell intelligens* The Centre for Working Life, 1989.
39. See PERBY, Maja-Lisa *Computerization and Skill in Local Weather Forecasting* in Göranzon, 1988, pp 107–142. The inner picture is a metaphor for professional knowledge in a special sense: how people with professional skills use their fund of knowledge to deal with a special situation. The employees at the Social Insurance Offices said that the inner picture is a good metaphor for their professional knowledge as well (see also Note 34).
40. See Note 21.
41. A basic component of the inner weather picture is that meteorologists form a picture of the weather as a continuous process of events in the atmosphere, and yet Maja-Lisa Perby chose to use the term 'weather picture' because the meteorologists' perception of the weather is tangible, pictorial.
42. The professional tradition referred to here is primarily one of cartographic analysis: the meteorologists repeatedly analyze the weather charts in the course of the day. This part of their work is a crucial aspect of the inner weather picture.
43. What appears to be particularly serious for the meteorologists in this involuntary exchange between different kinds of information is that the analysis of the charts has been made weaker – see the note above.
44. The meteorologists' long-term experience of such a sophisticated computerized aid as

numerical forecasts appears to be highly relevant in illustrating problematic aspects of the use of expert systems.

45. Maja-Lisa Perby discusses other conditions in work which favour a well founded inner weather picture.
46. WITTGENSTEIN, Ludwig *On Certainty*.
47. WITTGENSTEIN, Ludwig *Remarks on the Foundations of Mathematics* 3rd edn, Basil Blackwell, Oxford, 1978, p. 258.
48. See Note 26, Chapter 4 on John von Neumann and the letter from Lars Löfgren.
49. See JOHANNESSEN, Kjell S. *Rule Following and Intransitive Understanding*, published in Göranzon and Florin, 1990a, Chapter 5.
50. *Ibid.*
51. WITTGENSTEIN, Ludwig *Philosophical Investigations*.
52. *Ibid.* p. 117.
53. Josefson, 1983, pp. 45–51.
54. See this chapter, Notes 18, 19 and 20.
55. See the principle of sedimentation as a method in Chapter 2.
56. Wittgenstein, 1981.
57. See JANIK, Allan *Kompetens och expertis* in JANIK, Allan Cordelias tystuad, Carlssons, 1990.
58. TEMPTE, Thomas 1982, p. 60.
59. See, for example GORZ, Andre *Paths to Paradise*, 1984 and GÖRANZON, Bo *Gorz och datorernas tänkande*, Ord and Bild 1/85.
60. See, for example, NYBERG, Dan (Ed.) *Yrkesarbete i förändring* Carlssons, 1985, which describes the professional knowledge of aircraft mechanics.
61. Wittgenstein, 1978.
62. *Ibid.* p. 326.
63. *Ibid.* p. 320, p. 405.
64. See what Maja-Lisa Perby says (Notes 43 and 44) about the weakening of map analysis and maintaining the inner weather picture.
65. The term 'practical knowledge' comes from the Norwegian philosopher, Kjell S. Johannessen, Filosofisk Institutt, Bergen University.
66. See JANIK, Allan *Tacit Knowledge, Rule-Following and Learning*, and PRAWITZ, Dag *Tacit Knowledge – an Impediment for AI?* in Göranzon and Florin, 1990a.
67. *Ibid.*
68. See JOHANNESSEN, Kjell S. *Tankar om tyst kunskap* in Dialoger No. 6, 1988.
69. *Ibid.*
70. Wittgenstein, 1981.
71. See, for example the criticism in JANIK, Allan *Tacit Knowledge, Working Life and Scientific Method* in Göranzon and Josefson, 1988, pp. 53–66.
72. Compare Note 69. Johannessen will be publishing a comparative analysis of the meaning of the word in different philosophical works by, inter alia, Thomas Kuhn, Michael Polynai and Ludwig Wittgenstein, in a book entitled *Tyst kunskap – några huvudlinjer* which is to be published in 1992.
73. DEGERBLAD, Jan-Erik Planeringens vetenskapsteori, Projekteringsmetodik KTH, 1985.
74. Wittgenstein, 1981.
75. *Ibid.*
76. DEGERBLAD, Jan-Erik Yrkeskunskaper under svenskt 1700-tal- exemplet målaren och konservatorn, Erik Hallblad in Göranzon, 1988, pp. 80ff.
77. DREYFUS, Hubert L. and DREYFUS, Stuart E. *Mind over Machine: the power of human intuition and expertise in the era of the computer* Oxford, Basil Blackwell, 1986.
78. *Ibid.* pp. 105ff.
79. FAHLEN, Martin Persondatorn pa medicinkliniken in Göranzon, 1989 pp. 14–18.
80. PLATO, *The Republic*.
81. PLATO, *The Dialogues*. See Göranzon's editorial comment in Dialoger 1/Dialogens väsen. It was Pehr Sällström who drew my attention to this interpretation of Gorgias.
82. Göranzon et al. *Datorn som verktyg* Studentlitteratur, 1983, pp. 19ff.
83. When practically acquired knowledge dominates in an area, there is an inbuilt practice to articulate the experience in statements, formulated as conclusions of the experience. The motivation is the need to state reasons for recommendations and justification of viewpoints. This theme is discussed with regard to the role of the consultant in Göranzon, 1983, pp. 20ff.

Chapter 6

Education and Professional Knowledge

Computation and Analytical Skills

In just a few years time the method of calculation in current use will be replaced. Even if the basic principle of the calculation remains unchanged, there will be no question of computerizing current manual techniques. The new method will be designed with the requirements and potential of computer technology in mind. This will make the training more difficult. We do not yet know what the effects on professional knowledge will be.[1]

Before the advent of computers, roughly 75% of basic training for this work pertained to following the rules and assessing the material content of insurance coverage, with 25% being devoted to various administrative routines. When administration was computerized, these proportions were reversed. At the same time, new administrative routines were produced. It became increasingly important to concentrate on making sure that the right information was reported to the computer in the right way at the right time.[2]

These two statements were made at a conference on education in the future held under the auspices of the research project in September 1985.[3] The first statement is by Ulf Larsson of the National Board of Agriculture. It relates to the development of a new system for forest valuation (see Per Svensson's view of the long-term consequences of computerization in the caption). The conference was held in 1985, ten years after the review of the computer system had created a conflict at management level at the National Board of Agriculture.[4]

Håkan Svensson, who had followed the computerization of the Social Insurance Offices since the beginning of the 1960s, gives his view in the second quote and notes a radical change in occupational training in the area of social insurance work. The orientation towards the content of the work was replaced by training in technical skills. There is a link between this change and Andre Gorz' utopian description of future professional training that he outlines in his book *Paths to Paradise*.[5] In Gorz's picture of the future, the most important factor of working life is to acquire training in the new technology. He sees the microcomputer revolution as allowing people to change from one area of work to another with ease, once the technology has been mastered. The 'trivialization of work content' is important, because the value that our culture gives to work must be demolished and replaced by other values.[6]

The question of education in the future came to the fore when the dialogue computer was introduced at the end of the 1960s. This issue was highlighted in a series of seminars held at Johns Hopkins University in the Spring of 1970.

The physicist and head of research at the Pentagon, Eugene Fubini, thought that the users of new technology needed no practical knowledge of the work carried out by computers:

There is a fundamental difference between doing something and writing a general rule for doing it. The latter requires a higher degree of skill. Can you define in words the procedure followed in driving a car along a winding road or staying upright on a bicycle? You will soon find how different doing is from knowing how to do. I predict a steady decrease in the need to teach students how to do things that machines will do in their place, and a steady increase in the need to write general rules for doing these things.[7]

Fubini is yet another representative of the dream of the exact language. He felt that the need for training in calculation skills would disappear because such skills would be automated. The quote indicates what he imagined the future work of professional people would be, namely, formulating general rules or algorithms that would establish the procedure for an action they wished a computer to take. In the longer term, as machines learn more and more, this task would involve formulating goals and letting the computer find solutions.[8]

We may use debates in which different viewpoints are juxtaposed as a way of examining fundamental issues such as those involved in the process of professionally skilled people mastering their concepts and experiencing certainty in their actions and judgments. Several people who attended the series of seminars in the spring of 1970 at the Johns Hopkins University argued against Fubini. Thus, Patrick Suppes, Professor of Philosophy at Stanford, talked about learning mathematics.

There is not a single substantial scientific paper on how students learn any portion of advanced mathematics or the advanced portion of any other systematic science. It is certainly a matter of first-order scientific research to understand these matters better before we turn to a second-order strategy for education.[9]

Martin Shubik, Professor of Industrial Economics and Organization at Stanford University, made a brief contribution to the discussion, saying that 'we must restore apprenticeship training.'[10] This observation did not attract any comment in the discussion at the symposium.

In his book, *The Sciences of the Artificial*, Herbert Simon follows Eugene Fubini's line of thought. Simon says that professional groups need to learn to formulate generally applicable rules. In his opinion, formal techniques and methods from operations analysis should be part of the training of all technologists, sociologists, teachers and lawyers. Simon considers operational analysis to be a central discipline for every liberally educated man.[11]

In his book, *An Introduction to Mathematics*, published in 1966, the Swedish mathematician, Tord Ganelius, maintained that the new technology meant we could place less emphasis on routine work and exercises.[12] This reasoning has had a strong influence on the teaching of mathematics in Sweden over the last 15 years, manual calculations being replaced by the pocket calculator. In a comparative analysis among different OECD countries carried out in 1985, Swedish youngsters of fifteen to sixteen were well down on the scale of mathematical knowledge, while Japanese youngsters were at the top of the table.[13] The Japanese school system does not accept the use of technical aids before the basics have been mastered.[14]

The Swedish mathematician, Lennart Carleson, published a book entitled *Mathematics for our Time* as a contribution to the public debate. In dealing with the question of mathematical terms, Carleson writes that we give them a personal content. The abstract definitions must, of course, be understood but they are buried deep in our minds. It is primarily the idea and the function of the concept that is used when we work with these terms, and not the mathematical construction or the formal definition. The combination of formal definition and personal content is a central issue for Carleson.[15]

By contrast to Fubini and Ganelius, Lennart Carleson considers that the introduction of sophisticated technology in education means that we must emphasize the routine type of exercise. He maintains that one must accept that active and usable knowledge requires a great deal of routine training. It is unreasonable to think that an increase in the use of technology would mean that we could ignore routine manual work.[16] There are close links between the ability to calculate and the ability to analyze.

Carleson shares the same view as Wittgenstein about concepts and how one learns to master them.[17] In addition, he describes something of central importance in the case study of the County Agricultural Boards. In his statement that there are close links between the ability to calculate and the ability to analyze, he confirms the essence the forest rangers' constant references to manual calculation.

The Information Society

The term 'information society' was first used in 1972 in a Japanese futurological study carried out in anticipation of the revolutionary possibilities that the dialogue computer would open up. It was thought then that the information society would be a fact by the year 2000. The study concluded that the entire population needed training in computer technology as soon as possible while, if the vision of this study had become reality, a 'computer mind' should have been established by about 1985.

A different perspective emerges from a new Japanese futurological study carried out in 1985. The aim was still to introduce the information society, but there was no longer talk of the computer mind. On the contrary, the notion of the information society was placed in doubt when attention was focussed on questions such as the effects on people's self-confidence after prolonged dependence on computers:

I don't know why, but I feel depressed and lose my self-confidence. Human beings should use computers, but the machines are gradually taking over, and we are having to accommodate ourselves to their ways of working. Computers are like babies, they lack adaptability: they refuse to understand very human concepts like 'this is acceptable – let it pass'. They prefer black and white. I prefer grey.[18]

The study dwells on another aspect of the concept of certainty. The term 'functional autism' is used to describe the phenomenon of people who work for a long time in a computerized environment, with its characteristic categorization of reality into black/white or right/wrong, experiencing difficulties in confronting reality – a reality which is in essence the ability to cultivate social relationships.[19] In developing this line of reasoning, educational issues such as building on the long tradition of apprenticeship training in Japanese culture become important while computer technology training is no longer a priority.[20]

A worldwide network of researchers gave their views on the 1985 Japanese futurological study. Their findings were presented in a research report published in 1988.[21] The most striking aspect of their findings, seen from the perspective of this work, is that epistemological questions will become important in future research and education. They also emphasize the need in the development of science and technology for a balance to be struck between between technology and other knowledge, such as history and literature.[22]

Edward Feigenbaum and Pamela McCorduck are among those that state their firm faith in artificial intelligence – in particular in its most prominent application, the so-called 'expert systems'. In their book published in 1983, *The Fifth Generation. Artificial Intelligence and Japan's Computer Challenge to the World*,[23] Feigenbaum considers that with the help of expert systems we will be able to process all kinds of knowledge.

We have the opportunity at this moment to make a new version of Diderot's Encyclopedia, gathering up all knowledge, not just the academic kind, but the informal, experimental, heuristic kind.[24]

His partner, Pamela McCorduck, Professor of Literature, develops Feigenbaum's argument on expert systems from the perspective of the history of ideas in her book *Computer Culture* published in 1984.[25] She argues that expert systems pave the way to a democratization of knowledge for not only experts but all who is interested have access to sophisticated knowledge in a given area of activity, irrespective of their previous level of knowledge. She compares the spread of expert systems through computer technology with the Gutenburg printing press and the spread of the printed word. Quoting Feigenbaum, McCorduck maintains that expert systems will out-perform human levels of competence.[26]

Feigenbaum refers to Diderot and the French *Encyclopedia* project, but lacks the feeling for the complexity of professional knowledge expressed within the project.[27] In the tradition of the Enlightenment shared by Diderot, D'Alembert and Rousseau, there is a realization that no advanced dialogue can reject the kind of knowledge achieved only by specialist manual workers.[28] Expert knowledge is essential, but it is not all of one kind.[29]

At the conference on Culture, Language and Artificial Intelligence held in Stockholm in May–June 1988, the literary critic, Horace Engdahl, made a contribution on the information society where the distinction between the beginner and the expert was a crucial:

We have all learnt to realize that if the same book is put in the hands of an expert and a beginner, the amount of knowledge which the book contains is not the same for both. That is why we have teachers. Their presence is seen as an almost embarrassingly irrational element in our culture. The fact that teachers did not disappear when the art of writing was introduced gives us hope that we will gradually become less intoxicated by the prospect of artificial intelligence.[30]

In his contribution to the Stockholm conference Gerald Steig, a researcher in literature at the Sorbonne in Paris, sees the 'information society' as the final stage in a process of development. He cites the story of Golem and the master's apprentice, and the unfortunate consequences of confusing means with goals. Stieg captures the spirit of the times in the title of his article, 'I have no idea where I am going, so to make up for that I go faster.'[31]

There are many cliches about the phenomenon of professional knowledge in our information society. 'In only three weeks I gained three years' experience', said an advertisement from a leading computer supplier.[32] The problem with cliches, as Karl Kraus noted in his critical work carried out in Vienna around the turn of the century, was that they contained just enough truth to conceal the genuine falseness of the message.[33]

Education for the Practical Intellect

To put Feigenbaum and McCorduck's use of the word knowledge into some perspective, we have chosen to move back to the turn of the century, to a debate at the Stockholm Institute of Technology. This debate, which was documented by Bo Sundin and Nils Runeby, Professor of the History of Ideas, concerned the relationship between theory and practical work in the training of engineers as outlined in the writings of Johan Henrik Cederblom, Axel F. Enström and Johan Peter Klason.[34]

Johan Henrik Cederblom constantly refers in his writings to the position and tasks of engineers. In the preface to his work on steam engines published in 1889, Cederblom finds himself called upon to defend a textbook that is not merely a crib with tables, but which also explains the underlying principles involved. Cederblom considered that it was dangerous for engineers to apply formulae the origin of which they were totally ignorant. They stand on firmer ground if they are familiar enough with their field to need no other rules than those they can create themselves from their own knowledge. This is the first level of knowledge that makes it possible for engineers to develop for themselves the formulae they need for each particular task. Without this they are constantly exposed to the risk of error. The requirement is that one must be familiar enough with an area to be able to formulate rules and interpret them.[35]

Cederblom's views on the future professional knowledge of engineers echoes the tradition of ideas from a century earlier, from around 1800. Nils Runeby reproduces an idea from Johan Gottleib Fichte, in which the latter expresses a view of wisdom which takes the argument on the practical intellect a stage further. A person is only a man of learning when he has 'come through' the wisdom of his time and has formed it in his inner mind ('ausbilden'). Otherwise, one is simply 'something ambiguous, something in between'.[36] Runeby quotes Benjamin Höjer, who introduced Fichte's ideas in Sweden. Höjer had a variation on this theme, arguing in favour of not attempting to tie down language with rules.

If I can master my concepts, and not be mastered by them; if I am able to describe my system, to give an account of my concepts – then I am one of the initiated, an adept, a member of the Order of Philosophers, and have no need of the mentorship of others. I master them and the world.[37]

In a postscript to his article, Nils Runeby says that the view of knowledge developed here is not primarily a meeting between natural sciences, technology and the humanities. It concerns rather a more fundamental kind of understanding.[38] This viewpoint is central to the understanding of professional knowledge which we as members of the research project developed.

Let us return to the debate at the Stockholm Institute of Technology. Axel F. Enström, who was the first principal of the Academy of Engineering science at the turn of the century and for the next two decades, placed, like Cederblom, a strong emphasis on the importance of personal, practical activity in learning scientific theories, methodological rules and regulations. Reflecting on the knowledge derived from practical work, he wrote:

I maintain that what has appeared to be too much theory is instead too little theory, that is to say young engineers have a great deal of knowledge which is called theory, but this consists largely of rigid theory, dead knowledge. I have arrived at the definite opinion that our present scientific education is inadequate in the light of modern requirements. By science I do not mean what many people mean, the involvement with mathematics and formulae. . . . I would go so far as to say that a cobbler could be a scientist in his field. A person involved in practical work, an engineer or a cobbler, may, by reflecting on his work or his experience, achieve insights which can be compared with learning ready-made theories.[39]

A third example, but of a completely different kind, is the Nobel Institute chemist, Johan Peter Klason. Klason was interested in passing on a humanistic ideal in education and considered that it was important for engineers to study philosophy and literature, without which they would be liable to regard people at work as machines and would therefore find it difficult to act as managers. Klason's work was characterized by a strong interest in literature, the history of ideas and philosophy, while he also placed emphasis on in-depth competence as an engineer in one's own professional field.[40] Thus Klason proposed another aspect of reflection as a kind of knowledge for engineers.

Cederblom, Enström and Klason formulated their views about the knowledge of engineers at a time when computers did not exist. Today, similar reflections take place in the context of computer use.[41]

Mike Cooley, an engineer himself, became involved at an early stage in the significance of the design of computer support, and not least what the effects of such support were on the designers' professional knowledge.[42] He emphasizes what he calls the feeling of a designer or engineer for the physical world, and sees a danger in the increasing amount of abstraction involved in computer-supported design. In general, he sees certain risks in working more and more with models of reality.

If people spend more time working on models of reality rather than in direct contact with reality, and are thereby denied the learning process which is a result of this direct contact with one's senses, and an assimilation of tacit knowledge, then serious problems will occur over a longer period of time.[43]

Mike Cooley places a healthy question mark next to an often repeated idea, the idea that people and computers should interact to get the best from both sides.

When a person uses a machine, there is an interaction between two opposites. People are slow, inconsistent and unreliable, but have considerable creativity, while the machine is fast and reliable but totally uncreative. The starting point in designing programmes for computer-aided design is that these two inconsistent characteristics – the creative and the uncreative – are complementary and provide a good basis for a symbiosis between man and machine.[44]

Cooley observes that design method cannot separate these two elements and then combine them at a given time, 'as if it were a chemical process'. He emphasizes that this entire research area is a complex one, still poorly defined and with very meagre research findings.[45]

That we know far too little about the interaction between humans and computers is also apparent from the case studies. Good intentions are not enough when it comes to computer support. It is not enough to arrive at a theoretical solution for dividing work between people and computers. We need a more profound understanding of the practical intellect and respect the fact that there is more to professional knowledge than can be understood by an outsider.

Judging Light in Photography

One difficulty in approaching the practical intellect is that many contemporary jobs are outside our everyday experience. The valuations of forest rangers are one example. We have found it fruitful to consider more tangible or accessible activities that make it easier to discuss what happens in other jobs. The following reflections are the work of the photographer, Peter Gullers. His description of light in photography jogs the reader's memory of the physical impressions of different light conditions. It is easy to follow his reasoning and understand the problems created for photographers by the automation of light metering:

The text of a recent advertisement for cameras said: 'Instructions for taking good pictures – just press the button'. Thanks to new technology we no longer need to know a lot about the technique of photography in order to take good pictures. The manufacturer had built a programme into the camera, a programme which made all the important decisions and all the assessments needed to produce a satisfactory result.

New technology has made it easier to take photographs and photography has become very reliable and accurate in most normal conditions. When there is not enough light, the exposure is blocked or a built-in flash is activated to ensure satisfactory results.

The programme cannot be modified and no opinion can be formed of the results until later. The underlying principles are invisible – the process is a silent one. Neither does the manufacturer describe how the programme makes these assessments. In retrospect, when the picture has been developed, even the uninitiated judge can say that the picture is too dark, too light or blurred. On the other hand the cause of the fault is difficult to establish without a thorough knowledge of technology, or of the conditions under which the photograph was taken. There are numerous problem areas and the causes of these problems tend to merge with each other.

Physiologists claim that the eye is a poor light meter because the pupil automatically adapts to changes in the intensity of light. This may be so. When faced with a concrete situation that I have to assess, I observe a number of different factors that affect the quality of the light and thus the results of my photography. Is it summer or winter, is it morning or evening? Is the sun breaking through a screen of cloud or am I in semi-shadow under a leafy tree? Are parts of the subject in deep shadow and the rest in strong sunlight? Then I have to strike a balance between light and darkness. If I am in a smithy or in a rolling mill shop, I note how the light coming through the sloping skylights contrasts with the sooty heat of the air in the brick building. The vibrations from hammers and mills make the floor and the camera tremble, which makes photography more difficult and affects the light-metering. The daylight is enhanced by the red glow of the steel billets. In the same way I gather impressions from other situations and environments. In a new situation, I recall similar situations and environments that I have encountered previously. They act as comparisons and as association material and my previous perceptions, mistakes and experiences provide the basis for my judgment.

It is not only the memories of the actual process of photography that play a part. The hours spent in the darkroom developing the film, my curiosity about the results, the arduous work of recreating reality and the graphic world of the picture are also among my memories. A faulty assessment of the strength of the light and the contrast of the subject, the vibrations and tremors become important experience to be called upon next time I face a similar situation. All of these earlier memories and experiences that are stored away over the years only partly penetrate my consciousness when I make a judgment on the light conditions. The thumb and index finger of my right hand turn the camera's exposure knob to a setting that 'feels right', while my left hand adjusts the filter ring. This process is almost automatic.

The problem with automatic computer-aided light metering is that after a long period of use, one tends to lose one's ability to judge light conditions. Few people can manage without mechanical or electronic light meters today. But it is not simply the ability to judge light that is disappearing. Unless one regularly makes a manual judgment of light, one's sensitivity to shades of light tends to become blunted.

Our pictorial memories of past experiences are not activated in the same way unless they have been connected with similar assessments. Unless one regularly performs the actual work of producing pictures, the ability to make the best use of composition and light-modifying techniques when printing will wither too.

The problem with the automatic meter is not only that its program does not consider whether it is day or night, nor the nature of the subject, nor the inexperience of the user. The most important point is that it denies me access to my memories and blunts my perceptions and my ability to discern shades of light. This intimate knowledge is not linked to what I do when I photograph, that is, to the operations I perform, but to actual memories and experiences when I take photographs and when I develop and print pictures.[46]

Compare the work of a photographer in his darkroom with the calculations made by forest rangers! Both lead to more profound knowledge – in the dark room the photographer sees the results of how he has judged the light, while in the course of his calculations the forest ranger makes a more profound study based on his inventory of the forest. In both cases, knowledge is acquired at that point – and not before through routine work: the hours in the darkroom, the hours of manual calculation.

The sessions in the darkroom taken together, each of which reflects the time at which the photograph was taken, give an overview, an ability to judge, which the photographer takes with him when he is out taking photographs. The same applies to the forest ranger: when he has made his calculations of property after property following a series of inventories in the field, he takes his knowledge with him out into the forest when he next carries out a forest inventory.

Put another way, this could be called the feedback of results: a feedback between the photographs which emerge in the darkroom and the way the photographer makes his judgment of light conditions when he took the photographs or between the value which emerges when the forest ranger calculates and the inventory he takes in the forest. But, as can be seen from Peter Gullers' description, this feedback of the results to the practical intellect is something quite different, and far more genuine than a computer's rapid calculation of the results of various alternative input values.

Peter Gullers' description illustrates the concept of dialogue which has become a central aspect of the research method used. The background to the research project, the *Dialogue Seminar*, was a reaction to how words such as 'dialogue' and 'conversation' were eagerly bandied about in descriptions of the relationship between people and machines (and between machine and machine).[47]

Our perspective is that it is the diversity, the manifold nature of dialogue which is the point. We have been given an idea of the breadth of different meanings. It is inconsistency which gives the concept of dialogue its vitality, the paradox being that if we accept the meaning which confirms our preconceived notions, 'what we recognize', then we find ourselves in a different area – the area of the monologue.

The inner dialogue, which is closely related to what can be termed tacit knowledge, can take professional knowledge a step further by virtue of the fact that in work there is the opportunity to exchange ideas with colleagues. Dialogue is movement, discovering something together. It cannot be captured in a formula. Dialogue is a way of getting behind the formal aspects, and thus help us gain insight into professional skills.[48]

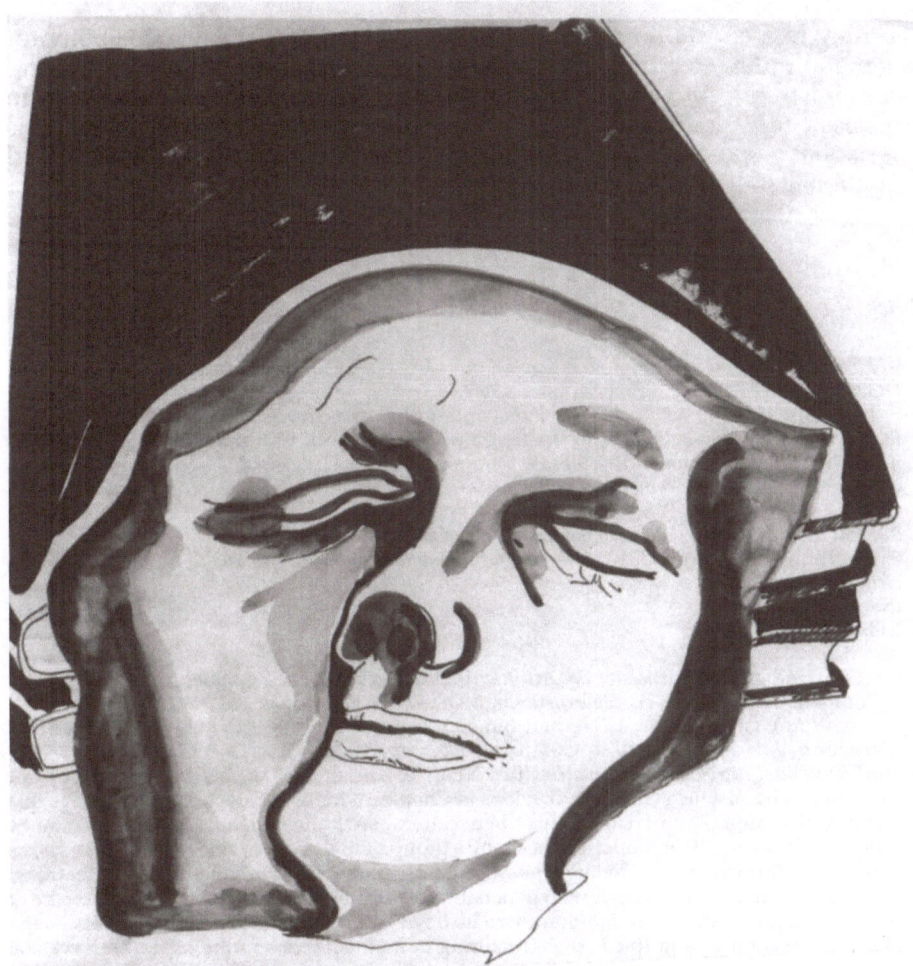

'*I*: If I were you, I would put that down on paper. It would be a pity to lose it.
HE: That's true, but you can't imagine how little I care about methods and rules. A person who needs rules will never get anywhere. Geniuses read very little, practise a great deal and create themselves.'

Denis Diderot, *Rameau's Nephew*
Illustration: Lennart Mörk

A Final Comment

When Wittgenstein was looking for a motto for his philosophical investigations, he considered using a quote from Shakespeare's *King Lear*: 'I will teach you differences'.[49] It is in this spirit – in the interest of the manifold nature of reality and its complex logic – that I have attempted to examine the theme of the use of computers and professional knowledge. In other words, the case studies, which focus on the ability to calculate, act as a mirror in which we 'recognize ourselves' and form the basis for comparison with our own activities. A person who has never seen the differences will never understand the similarities.[50]

Turning to a question as vital as an occupational group's responsibility for the general public, we have seen similarities between the forest rangers and the administrative and executive staff manning the social insurance offices. In both cases, the employees have stated that their ability to assume this professional responsibility was linked to well developed and well maintained occupational skills.

Notes

1. LARSSON, Ulf: Presentation vid samtal om effekter av datoranvändning, in Göranzon: *Utbildning-Arbete-Teknik*: Arbetsrapport, 1985.
2. SVENSSON, Håkan: ADP och yrkeskunskaper i försäkringskassan, *Ibid.* p. 35.
3. *Ibid.* The conference was held in cooperation with the Council of Nordic Ministers for people in the educational sector in the Nordic bloc.
4. Compare Chapter 2, *A growing conflict.*
5. GORZ, Andre *Paths to Paradise*, 1984.
6. See GÖRANZON, Bo: *Gorz och datorernas tänkande*, Ord & Bild 1/85.
7. GREENBERGER, Martin (Ed.): *Computers, Communications and the Public Interest* The Johns Hopkins Press, 1971, p. 129.
8. *Ibid.* p. 130.
9. *Ibid.* p. 131.
10. *Ibid.* p. 134.
11. SIMON, Herbert *The Sciences of the Artificial* MIT Press paperback edition, 1970.
    Compare GEORGE, F. H.: *Cybernetics in Management* Pan Books Ltd., 1970, p. 5.
    Tore Nordenstam discusses Herbert Simon's view of education the chapter entitled *Två oförenliga traditioner* published in Göranzon: 1983, pp. 51–53.
12. Tord Ganelius: *Introduktion till matematiken* Natur & Kultur, 1966, p. 242. There is a polemic strain in this book which Ganelius develops in Chapter 1, Matematik och kultur. He says that many of the attempts to create contact between material and humanistic cultures may be compared to 'a display of modern dancing by a group of hydraulic engineers, that is to say, as attractive as it is clumsy'. Ganelius argues for an orientation towards the history of learning. Its development and interaction with general trends and intellectual growth appears to be of considerable interest. 'In my opinion there is, however, a more central connection between the separate cultures and this is the structuring of mathematics and the natural sciences in, first and foremost, education to avoid their eclipsing the humanistic values' (p. 21). Undeniably, this subject is due for more profound reflection – the meeting of the symbols on the Stockholm Institute of Technology's emblem: Art and Science. See the introduction to Göranzon (Ed.), 1978.
13. See the concluding documented in the article, Vad ska vi göra åt matematikundervisning? in *Matematik i Skolan*, No. 4, 85/86, and The Ministry of Education: *Matematik i skolan. Översyn om undervisningen i matematik inom skolväsendet* DsU 1986: 5.

14. CARLESON, Lennart: *Matematik för vår tid* Prisma, 1968.
    Carleson criticizes what he calls a fussy and detailed interest in stringency which remains superficial and which runs the risk of making the subject boring and uninteresting. As an example, he takes the theory of sex, which may be made difficult by using complicated terminology and many symbols, 'which seems to be the main point of many presentations. In fact it is a matter of ordinary common sense and is just as simple or perhaps even simpler than 1 + 1 = 2. There is a considerable risk that by presenting what is obvious and well-known in a complicated way, one's work runs counter to one's objectives.'
15. *Ibid.* p. 134.
16. *Ibid.* p. 128.
17. See, for example, Wittgenstein, 1978, pp. 199–203.
18. National Institute for Research Advancement (NIRA): *Comprehensive Study of Microelectronics* 1985, p. 35.
19. *Ibid.* p. 28.
20. *Ibid.* p. 29.
    Bill Ford: *A Learning Society. Japan through Australian Eyes* In Magnus Florin and Bo Göranzon (Eds.) 1990a.
21. NIRA: *A Comprehensive Study of Microelectronics: Views of 100 Wise Men around the World*, Vol. 1, No. 2, 1988.
22. *Ibid.* p. 35.
23. Edward Feigenbaum and Pamela McCorduck are among those that state their firm faith in artificial intelligence – in particular in its most prominent application, the so-called 'expert systems'. In their book *The Fifth Generation. Artificial Intelligence and Japan's Computer Challenge to the World*, 1983.
24. *Ibid.* pp. 49ff.
25. See McCORDUCK, Pamela: *Knowledge Technology: The Promise* in Pagels (Ed.), 1984, p. 108.
26. *Ibid.* pp. 108ff.
27. Feigenbaum & McCorduck, 1983, pp. 49ff.
28. See ENGDAHL, Horace *Dialog och Upplysning* in Dialoger 3/86.
29. See BERGENDAL, Gummar (Ed.): *Information, Kunskap, Ansvar* Studentlitteratur, 1987.
30. ENGDAHL, Horace *The Personal Signature* in Göranzon & Florin, 1990a.
31. STEIG, Gerald 'I have no idea where I am going, so to make up for that I go faster', in Göranzon & Florin, 1990a.
32. Göranzon: *Estetik och teknik i datorsamhället*, in *Digitalteknik – konstnärlig uttryck* Konstindustriskolan, The University of Gothenburg.
33. JANIK, Allan *Humanvetenskapens kardinalproblemoenigheten om begreppens innebörd* published in Bo Göranzon: (Ed.): *Den Inre Bilden* Carlssons, 1988, pp. 37ff.
34. RUNEBY, Nils *Större fart framåt – Kring en injenjörs föreställningsvärld* published in Bo Sundin *I teknikens bakspegel*, an anthology of the history of technology, Carlssons, 1987, and Bo Sundin: *Injenjörsvetenskapens tidevarv* Almqusit & Wiksell International, 1981.
35. *Ibid.* p. 320.
36. RUNEBY, Nils *Om självmedvetandets rätt* published in JACOBSSON, Per (Ed.): *Injenjörerna, Vetenskapen och Värderingar*, a series of seminars at the Stockholm Institute of Technology, Spring 1988, p. 3.
37. *Ibid.* p. 7.
38. *Ibid.* p. 8.
39. Sundin, 1981, p. 91.
40. *Ibid.* p. 140.
41. See HULT, Jan *Vad har hänt med räknestickan + Ingenjörsarbete i förändring* in Sundin, 1987.
42. Mike Cooley, 1973.
43. Mike Cooley, 1987, p. 34.
44. *Ibid.* p. 38.
45. *Ibid.* p. 39.
46. GULLERS, Peter *Ljusbedömning i fotografi* in Göranzon, 1983, pp. 31ff.
47. See Bo Göranzon, editorial comment in Dialoger 4/87, *Den oavslutade dialogen.*
48. SÄLLSTRÖM, Pehr *The Essence of Dialogue* in Göranzon & Florin, 1990b.
49. EAGLETON, Terry *Wittgenstein's Friends* in *The New Left Review*, No. 13, p. 64.
    Rush Rhees: *Recollections of Wittgenstein* OUP, p. 64.
50. ZERN, Leif *Venus armar* Nordstedts, 1989.

References

ANDERSSON, John *Industriföretagets produktionsstyrning* Ind Ek Org, KTH, 1974
ARISTOTLE *The Nichomachean Ethics* Oxford University Press, 1984
ARNSTBERG, Karl-Olov *En etnologisk studie av yrkeskompetensen på ett försäkringskassekontor* Stockholm, Centre for Working Life, 1989
  – *ADB inom den allmänna försäkringen – på 1980-talet och därefter* Interim Report Ds 1979:4, Stockholm, Centre for Working Life
  – *Socialförsäkringens datorer* SOU 1981:24, Stockholm, Centre for Working Life
ÅGE, Per-Johan *Skogsvärdering. Översiktlig beskrivning av skogsvärderingsproceduren och dess utveckling, med särskild vikt på uppdelning i arbetsmoment* (manuscript) Lantbruksstyrelsen, 1975
BATESON, Gregory *Steps to an Ecology of Mind* Chaldler Publishing Company, 1972
BOLTER, David *Turing's Man. Western Culture in the Computer Age*, London, Duckworth, 1984
BOOLE, George *Matematisk analys av logik* Sigma Vol. 5, Forum, 1960
BUTTIMER, Anne *Creativity and Context* Lund's Studies in Geography, Human Geography No. 50, The Royal University of Lund, Department of Geography, 1983
CARLESON, Lennart *Matematik för vår tid* Prisma, 1968
CHEKHOV, Anton *Three Sisters*
CHURCHMANN, C. W. *The Systems Approach* N.Y. 1966
COBB, Noel *Prospero's Island* Coventure Limited, 1984
COOLEY, Mike *Computer Aided Design – its nature and implications* TAAS Publication, 1973
  – *Architect or bee?* London, The Hogarth Press, 1987
  – d'ALEMBERT *Inledning till Encyklopedin* translated by Jan Stolpe, Uppsala, Carmina klassiker, 1981
DAHL, Eva-Lena *Överideologi och politisk handlingsprogram. En studie i Lockes och Rousseaus tänkande* Göteborg, Acta Universitatis Gothoburgensis, 1980
  – *Synen på kunskap i Jean-Jacques Rousseau's ideologi*, in *Dialoger 1/86*
DAHLSTRÖM, Edmund, GARDELL, Bertil, and RUNDBLAD, Bengt G. *Teknisk förändring och arbetsanpassning* Prisma, 1966
DANIELSSON Albert *Företagsekonomi – en översikt* Studentlitteratur, 1975
  – *et al. Samtal om Ledarskap, Ledning & Ledare* Svenska Dagbladet, 1986
  – *Teknik-Människa-Samhälle*
de MONTAIGNE, Michel a presentation in *Idéhistorisk läsebok*, Vol. 1-Gidlunds, 1982
DEGERBLAD, Jan-Erik *Planeringens vetenskapsteori Projekteringsmetodik*, KTH, 1985
  – *Yrkeskunskaper under svenskt 1700-tal-exemplet målaren och konservatorn, Erik Hallblad*, in GÖRANZON, 1988
DENETT, Daniel The Role of the Computer Metaphor in Understanding the Mind in Heinz R. Pagel Computer Culture. The Scientific, Intellectual and Social Impact of the Computer, *Annals of the New York Academy of Sciences*, Volume 426, New York 1984
DESCARTES, René Avhandling om metoden, *Idéhistoriks läsebok*, Vol. 1, Gidlunds, 1982
DIALOGER 1/86 DIALOGENS VÄSEN
  – *2/86 Den inre dialogen*
  – *3/86 Dialog och upplysning*
  – *4/87 Den oavslutade dialogen*
  – *5/87 Artificiell Intelligens*
  – *6/88 Tyst kunskap*
  – *7–8/88 Artificial Stupidity*

- 9/88 *Den andre*
- 10/88 *Datorer och kunskap*
- 11–12/89 *Arbetets värde*
- 13/89 *Encyklopedi*

DIDEROT, Denis *Brev till Sophie Valland* translated into Swedish by Olof Nordberg, Atlantis, 1987
- *Skådespelaren och hans roll* Prisma, 1963
- *Rameus brorson* translated into Swedish by Ria Wagner, Tiden, Stockholm, 1951
- *Instinkt* translated into Swedish by Jan Stolpe, in *Dialoger*, No. 13/89

DREYFUS, Hubert L. *What Machines Can't Do: the Limits of Artificial Intelligence* Harper Colophon Books, 1979
- and DREYFUS, Stuart E. *Mind over Machine. The Power of Human Intuition and Expertise in the Era of the Computer* Oxford, Basil Blackwell, 1986

EAGLETON, Terry Wittgenstein's Friends in *The New Left Review*, No. 135, September/October 1982

ECKERBOM, Gertie, *De framtida yrkeskunskaperna – några synpunkter från försökskontoren* in Bo Göranzon and Reidar Bergström (Ed.) *De framtida yrkeskunskaperna. En arbetsbok för reflektion* The National Social Insurance Board, 1990.

EDSTRÖM, Anders in cooperation with Andersson, Björn and Ohlsson, Bengt *Terminaler och styrsystem* Företagsekonomiska studier, The University of Gothenburg, 1972:15

EDSTRÖM, Olof *Man-Computer Decision Making: the development of three terminal systems for empirical research*, Gothenburg's Studies in Business Administration, 1973:20

EK, Sverker R. *Spelplatsens magi. Alf Sjöbergs regikonst* Nordstedts, 1988

EKERWALD, Karl-Göran *Diogenes lykta* Rabeén and Sjögren, 1983

ENGDAHL, Horace *Dialog och upplysning* in *Dialoger* 3/86, Dialog och upplysning.
- The Personal Signature in GÖRANZON and FLORIN, 1990a

ERIKSSON, Gunnar, FRÄNGSMYR, Tore and von PLATEN, Magnus *Vetenskapens Träd* Stockholm, 1974

FAHLÉN, Martin *Persondatorn på medicinkliniken* in GÖRANZON, 1989

FEIGENBAUM, Edward and McCORDUCK, Pamela *The Fifth Generation. Artificial Intelligence and Japan's Computer Challenge to the World* 1983

FLORIN, Magnus *Skill and Technology* The Centre for Working Life, 1991
- *Michelangelos kupol*, in *Dialoger* 13/89

FOA OPERATIONSANALYS *FOA orienterar OM*, No. 8, 1967

FORD, Bill *A Learning Society: Japan through Australian Eyes*, in FLORIN and GÖRANZON (Eds.) 1990a

FORSBERG, Gun-Marie *Operatörsrollen för en datorbaserad modell för skogsvärdering* in GÖRANZON, 1978a, p. 147

FROSTENSSON, Katarina *Språket och den andra*, in *Dialoger* 9/1988

FRÄNGSMYR, Tore *Drömmen om det exakta språket* in ERIKSSON, Frängsmyr and von PLATEN *Vetenskapens träd*, Stockholm 1974

FÜRTH, Thomas *Organizationssyn och teknikval inom offentlig administration 1930–1980. Exemplet socialforsäkringen* in GÖRANZON, 1983

GANELIUS, Tord *Introduktion till matematiken* Natur & Kultur, 1966

GEORGE, F.H. *Cybernetics in Management* Pan Books Ltd., 1970

GORZ, Andre *Paths to Paradise* 1984

GREENBERGER, Martin (Ed.) *Computers, Communications and the Public Interest* The Johns Hopkins Press, 1971

GUILBAUD, G. T. H. *Cybernetik* Aldus/Bonniers, 1962

GUILLET de MONTHOUX, Pierre and GÖRANZON, Bo *Beskrivningsaxlar för informationssystem* Ind EK ORG, KTH, 1973a (arbetsrapport)

GULLERS, Peter *Verktygsmakare och operatörer* The Centre for Working Life, Report 37, 1982
- *Ljusbedömning i fotografi*, in GÖRANZON, 1983

GÖDEL, Kurt *Collected Works* vol. 1, Publications 1929–1936, Oxford University Press, 1986

GÖRANZON, Bo *Tabeller över lösningar till Lanchestersekvationen* FOA P-rapport, C 8164–2, 1967
- *Administrativt systemarbete* Armeéstabens OA-rapport, C2, 1969
- *Perspektiv på systemutvecklingsprocessen* (The Systems Development Process – a Perspective) SINFDOK, 1974b
- *The Annual Conference of the Operational Research Society* Torbay, England, SINFDOK, 1973
- *Arbetstagarnas ansvar och interesse i den tekniska utvecklingen* (two part) SINFDOK 1974a
- *Konstens pedagogiska funktion – några exempel* in GÖRANZON, JONSSON and MELBERG, 1978c

- *Att se Calibanmetaforen i vår teknologiska kultur* in *Dialoger* 9/1988 Den andre
- *et al. Datorn som verktyg* Studentlitteratur, 1983a
- *Turing's möte med Wittgenstein* in *Dialoger* 5/87 Artificiell Intelligens
- och SANDEWALL, Erik *Datorn – herre eller slav?* Forskningsrådsnämnden, Source No. 13, 1981
- *Verksamhetslust och bildning* in GÖRANZON, 1983b
- and FLORIN, Magnus, (eds.) *Dialogue and Technology. Art and Technology* Springer-Verlag, London, 1990a
- (Ed.) *Ideologi och systemutveckling* (A contribution to the discussion on science, technology and society). Studentlitteratur, 1978b
- (Ed.) *Datorutvecklingens Filosofi. Tyst kunskap och ny teknik* Carlssons 1983b
- and Bergström, Reidar (eds.) *De framtida yrkeskunskaperna. En arbetsbok för reflektion* The National Social Insurance Board, 1990
- Jonsson, Inge and Melberg, Arne (eds.) *Konst och samhällsförändring* An interview in *Sigtuna* 7–8 October 1977, Report No. 10, SALFO/FRN, 1978c
- *et al. Perspectiv på data system utveckling. Om data tekniska milöer, systemutvecklingsprocessen och arbetsorganisation* Studentlitteratur, 1978a
- *et al. Job Design and Automation in Sweden. Skills and Computerization* Report No. 36, The Centre For Working Life, 1982
- and JOSEFSON, Ingela *Knowledge, kill and Artificial Intelligence* Springer Verlag, London, 1988a
HART, Anna *Knowledge Acquisition for Expert Systems* in GÖRANZON and JOSEFSON *Knowledge, Skill and Artificial Intelligence* Springer-Verlag, London 1988
HARTZELL, Svante *Honeywell Bull Time-Sharing Service – a broad presentation* (information brochure) 1971
HEDBERG, Bo, SJÖBERG, Sam and TARGAMA, Axel *Styrsystem och företagsdemokrati* Företagsekonomiska studier, Gothenburg University, 1971:14
HEDBERG, Bo *On man-computer interaction: an organizational behavioural approach* BAS 1970:6, Göteborg, 1970
- *Dialogdatorn – sjuttiotalets räknesticka?* (The dialogue computer – the slide rule of the seventies?) STF engineers training course, 1973, Proceedings
HILTON, Julian *Teater och teknologi: Pygmalion och myten om den intelligenta maskinen* in *Dialoger* 6/8, Tyst kunskap
HODGES, Andrew and TURING, Alan *The Enigma of Intelligence* Counterpoint, Unwin Paperbacks 1983
- *Turing's Conception of Intelligence* in GREGORY and MARSTRAND *Creative Intelligences* London, Francis Pinter Publishers, 1987
HOYER, Rolf *Over till EDB. Undersokelser og vurderinger av databehandlingens innvirkning på organisasjon och miljo* Tanum Forlag, 1974
HYMAN, Anthony *Charles Babbage: Pioneer of the Computer* Oxford University Press
HÅFSTRÖM, Jan *Praktiken i måleriet* in *KRIS* No. 25–26, 1983
IVA *Important Technological Trends. Artificial Intelligence and Computer Science* IVA Report 246, Stockholm 1983
JANIK, Allan and TOULMIN, Stephen *Wittgenstein's Vienna* Simon and Schuster, New York, 1973
JANIK, Allan *Offenbach – konsten mellan monolog och dialog* in *Dialoger* 4/87, Dialog och upplysning
- *Humanvetenskapens kardinalproblem-oenigheten om begreppens innebörd* in GÖRANZON (Ed.) *Den Inre Bilden* Carlssons, 1988a
- *Tacit knowledge, Working Life and Scientific Method* in GÖRANZON and JOSEFSON, 1988b
- *Style, Politics and the Future of Philosophy* Kluwer Academic Publishers, Dordrecht, 1989
- *Tacit Knowledge, Rule-Following and Learning* in GÖRANZON and FLORIN, 1990a
- *The Role of Literature in the Theory of Knowledge* in GÖRANZON and FLORIN, 1990b
- *Caliban's Revenge* in GÖRANZON and FLORIN 1990c
- *Kompetens och expertis* in JANIK, Cordelias Tystuard, Carlssons, 1990
JAPAN COMPUTER USERS' DEVELOPMENT INSTITUTE *The Plan for the Information Society – a National Goal Toward Year 2000* Computerization Committee, Final Report, May 1972
JOHANNESSEN, Kjell S. *Tradisjoner og skoler i moderne vitenskapsfilosofi* Sigma Forlag A.S., 1985
- *Tankar om tyst kunskap* in *Dialoger* No. 6/1988.
- *Rule Following and Intransitive Understanding,* in GÖRANZON and FLORIN, 1990a
JOSEFSON, Eva-Karin *Människan och tekniken i litteraturen. En idéhistorisk studie om arbetets värde med exempel från två perioder*

JOSEFSON, Ingela and GULLERS, Peter *Begripa och förstå. Forskning om metoder att förmedla resultat i arbetslivsforskningen* The Swedish Centre for Working Life, Report No. 8, 1983
   – (Ed.) *Språk och erfarenhet* Carlssons, 1985
   – *Från lärling till mästare* FOU Report 25, SHSTF/Studentlitteratur, 1988
   – *Language and Experience* in GÖRANZON and FLORIN (eds.) 1990a JOSEPH, Herbert *Diderot's Dialogue of Language and Gestures* Ohio University Press
KARLQUIST, Anders *Om skapande improvisation – några reflektioner utifrån matematikens perspektiv* in SÄLLSTRÖM, (ed): 1984
KLINE, Morris *Matematiken i den västerlänska kulturen* Prisma, 1968
KLOCKARE, Barbara *Report from a visit to ALVEY Demonstrator Project* Agency for Administrative Development, 1986–06–11
KUHN, Thomas *The Structure of Scientific Revolutions* 2nd edition, Chicago, 1969
LARSEN, Steen *Den arbetande hjärnan. Sammanhanget mellan srbetets organisation och hjärnans funktion* Prisma, 1982
LAURIKAINEN, K.V. *Vetenskapens möjlighet och dess gränser, in Göranzon (ed), Den Inre Bilden* Carlssons Bokförlag, 1988,
LINDBORG, Rolf *Maskinen, människan och doktor La Mettrie* Doxa, 1983
LINDGREN, Michael *Glory and Failure* Linköping Studies in Arts and Science, 1987
   – *Dator för 150 år sedan – En historia om ett misslyckande* in SUNDIN (Ed.): *I Teknikens Backspegel. Antologi i teknikhistoria* (In the Rearview Mirror of Technology, an anthology of the history of technology), Carlsson Bokförlag, 1987, p. 233.
MACH, Ernst *Vetenskapens ekonomi* Sigma, Band 5, Forum 1960
MASON, John *The Irresistible Diderot* Quartet Books, London, 1982,
MELBERG, Arne *Estetisk verkan – några problem* in GÖRANZON, JONSSON and MELBERG, 1978b
MOLANDER, Bengt *Acting with Good Reason. Knowledge and Ignorance in Human Action* UHÄ Newsletter 1989:1
   – Socratic Dialogue: On Dialogue and Discussion in the Formation of Knowledge in GÖRANZON and FLORIN, 1990a
NATIONAL INSTITUTE FOR RESEARCH ADVANCEMENT (NIRA) *Comprehensive Study of Microelectronics*, 1985
   – *Views of 100 Wise Men around the World on 'A Comprehensive Study of Microelectronics, 1985'* Vol. 1 No. 2, 1988
NATIONAL SOCIAL INSURANCE BOARD, Sweden *Research assignment on the long-term orientation of ADP work in the social insurance system etc.* Draft Report 1986–09-xx
   – *Request for appropriation for the development of a decision support system in the field of social insurance* 1985–10–6
Von NEUMANN, John *A General and Logical Theory for Automatons* (Swedish translation in) *Sigma*, vol. 6, Forum, 1960
   – *The Theory of Selfreproducing Automata* edited and completed by Arthur Burks. Urbana and London, University of Illinois Press, 1966
NIGHTINGALE, Florence *Notes on Nursing, What is and what it is not* London, 1859, (new reprint, Duckworth, 1970)
NORDENSTAM, Tore *Sudanese Ethics* Almquist & Wiksell, 1975
   – *Värderingar och paradigm vid datasystemutveckling. Exemplet ALLFA utredningen* The Swedish Centre for Working Life, 1980
   – *Två oförenliga traditioner* in GÖRANZON, 1983
NYBERG, Dan (Ed.) *Yrkesarbete i förändring* Carlssons, 1985
OLSSON, Anders *Blindhet, begär och negativ utopi i King Lear* KRIS No. 17 /18, 1981
   – *Den okända texten*
PAGELS, Heinz R. Computer Culture. *The Scientific, Intellectual and ocial Impact of the Computer* Annals of the New York Academy of Sciences, Vol. 426, New York, 1984
PERBY, Maja-Lisa *Den inre väderbilden – teknikbedömning frånett mitt-i-arbetet-perspektiv* in GÖRANZON, 1988
   – Computerization and Skill in Local Weather Forecasting in GÖRANZON and JOSEFSON 1988a
   – The Inner Weather Picture in GÖRANZON and FLORIN, 1990a
PLATO *Dialogues* PLEJEL, Agneta *Vindspejare* Nordstedts, 1987
   – *Arbetet, Tjekov och Bibeln* in *Dialoger* 11–12, 1989
PRATT, Vernon *Thinking Machines: the Evolution of Artificial Intelligence* Oxford, Basil Blackwell
PRAWITZ, Dag *Tacit knowledge – an Impediment for AI?* in GÖRANZON and FLORIN, 1990a

PRINTZ-PAULSON, Göran *Turingmaskin* in the anthology of poetry *Säg, Minns Du Skeppet Refanaut?* Bonniers, 1984,

RHEES, Rush *Recollections of Wittgenstein*

ROSENBLUTH, Arthur, WEINER, Norbert and BIGELOW, Julien *Behaviour, Purpose and Teleology* in *Philosophy of Science*, 10, 1943

ROUSSEAU, Jean-Jacques *Emile* J.M. Dent & Sons, Ltd., 1982

ROYAL DRAMATIC THEATRE, Stockholm *Stormen* (The Tempest) Programme, 1968

RUNEBY, Nils *Teknikerna, Vetenskapen och Kulturen* Uppsala, 1977

– *Större fart framåt – Kring en injenjörs förestäl lningsvärld* in SUNDIN *I teknikens bakspegel* (an anthology of the history of technology) Carlssons, 1987

– *Om självmedvetandets rätt* in Per Jacobsson (Ed.) *Injenjörerna, Vetenskapen och Värderingar* 1988

SANDBERG, Åke *Perspektiv på organisationers beslutsfattande* in Göranzon, 1972

SCHOPENHAUER, Artur *The World as Will and Representation* (two volumes) Doven Publications, 1969

SEARLE, John *Kognitivism och datormetaforer* in *Dialoger*, No. 7/8, Artificial Stupidity, 1988

SHAKESPEARE, William *The Tempest*

SIMON, Herbert *The Sciences of the Artificial* MIT Press, 1970

SJÖBERG, Alf *Galilei och forskningens frihet* in GÖRANZON 1978a

– *Ögats roll (Troilus and Cressida)* in SJÖBERG, 1982

– *Teater som besvärjelse* Nordstedts, 1983.

SOCIAL INSURANCE EMPLOYEES' AND INSURANCE AGENTS' UNION *DATORN: Studie – och informationsmaterial om datoranvändning i försäkringskassorna* The Swedish Correspondence School, 1979

– *DATORN:* Studiecirkelansvar, 1980a

– *ADB inom försäkringskassorna* Action programme for the Social Insurance Employees' and Insurance Agents' Union, 1980b

STEIN, Dorothy *Think Tanks* in *The Guardian*, 26th March 1987

STOCOYEV, Vladimir *Mozart och Salieri* in *Dialoger* 5/87, Artificiell Intelligens

SUNDIN, Bo (Ed.) *Is the Computer a Tool?* Almqvist & Wiksell, Stockholm, 1980

– *Injenjörsvetenskapens tidevarv* Almqvsit & Wiksell International, 1981

– *I teknikens bakspegel* (an anthology of the history of technology) Carlssons, 1987

SVENSSON, Håkan *ADP och yrkeskunskaper i försäkringskassan*

SVENSSON, Per *Tjänstemannaansvar* unpublished manuscript, 1976–10–10

– *Systemutvecklingsprocessen för en datorbaserade modell för ekonomisk kalkylering av skog* in GÖRANZON, 1977b

– *Systemutvecklingsprocessen för en datorbaserade modell för skogsvärdering* in GÖRANZON, 1978a

– *Om expertsystem – att veta eller att mäta* 1987–12–14, unpublished manuscript

SWEDISH AGENCY FOR ADMINISTRATIVE DEVELOPMENT *A sub-project on the information system FF. A report of a series of interviews*, PM 1986, Stockholm, Sweden

SWEDISH CENTRAL ORGANIZATION OF SALARIED EMPLOYEES (TCO) *'Angående Teknik-värdering'*, TCOs ingenjörs- och naturvetarråd, 1974

SWEDISH MINISTRY OF EDUCATION AND CULTURAL AFFAIRS *Matematik i skolan. Översyn av undervisning i matematiken inom skolväsendet* DsU: 1986:5

SWEDISH MINISTRY OF HEALTH AND SOCIAL AFFAIRS, Stockholm, Sweden

– ALLFA research project *The committee's terms of reference*, Dir. 1977:52

– ALLFA project *Socialförsäkringens datorer* SOU 1981:24

– ALLFA project *ADB inom den allmänna försäkringen – på 1980-talet och därefter* Interim Report DS 1979:4

SWEDISH WORK ENVIRONMENT FUND *Systemutveckling. Pesentation av fyra olika sysnsätt* (Development Programme for New Technology, Work Organization and the Working Environment) The Swedish Work Environment Fund, 1984

SÄLLSTRÖM, Pehr (Ed.) *Diderot, Goethe och naturvetenskapen* in *Dialoger* No. 13/89

– *Matematisk notation* SALFO/FRN, 1989

– *Diderot, Goethe och naturvetenskapen* in Dialoger 13/89

TEMPTE, Thomas *Arbetets Åra. Om hantverk, arbete. Några rekonstruerade verktyg och maskiner* The Swedish Centre for Working Life, 1982

– *The Chair of Tut anch Amon* in GÖRANZON and FLORIN *Dialogue and Technology: the Working Memory* Springer Verlag, London, 1990b

TOBIN, N. R. and BUTFIELD, T. E. *OR Branch – British European Airways* Proceedings of IFORS Conference, August, 1972, Svenska Operationsanalysföreningen.

TORSTENSSON Lennart *Utredaretik i statsförvaltningen* in GÖRANZON 1978a

TOULMIN, Stephen *The Dream of an Exact Language* in GÖRANZON and FLORIN, 1990b

TURING, A.M. *On Computable Numbers, with an application to the Entscheidungsproblem* London Math. Soc. (2), 42. 1937
  – Computing Machinery and Intelligence *Mind*, October 1950

Von NEUMANN, John *A General and Logical Theory for Automatons* (Swedish translation in) *Sigma*, vol. 6, Forum, 1960
  – *The Theory of Selfreproducing Automata* edited and completed by Arthur Burks. Urbana and London, University of Illinois Press, 1966

WEINER, Norbert *Cybernetics, or control and communication in the animal and machine* M.I.T Press and John Wiley & Sons, Inc. 1961
  – *Materia, Maskiner, Människor. Cybernetiken och Samhället*, Forum 1952

WEIZENBAUM, Josef *Computer Power and Human Reason. From judgement to calculation* W. H. Freeman and Company, San Francisco, 1967

WHITEMORE, Hugh *Breaking the Code* Amber Lane Press, 1987.
  – *Enigmakoden* The Royal Dramatic Theatre, Stockholm, 1988 (Translated into Swedish by Per-Erik Wahlund)

WINOGRAD, Terry and FLORES, Fernando *Understanding Computers and Cognition. A new foundation of design* Ablex Publishing Company, 1986

WITTGENSTEIN, Ludwig *Remarks on the Foundations of Mathematics* 3rd edn Oxford, Basil Blackwell, 1978
  – *Philosophical Investigations*
  – *On Certainty*

von WRIGHT, Georg Henrik *Logik, filosofi och språk* Aldus, 1971

YATES, Frances *Shakespeare's Last Plays* London, Routledge & Kegan Paul, 1975
  – *The Art of Memory.* London, Ark paperbacks, 1984

ZERN, Leif *Älskaren och mördaren. Shakespeare och den andra spelplatsen* Alba, 1984
  – *Venus armar* Nordstedts, 1989

Name Index